The Politics of Fresh Water

Water scarcity is not simply the result of what nature has to offer but always involves power relations and political decisions. This volume discusses the politics of the freshwater crisis, specifically how access to water is determined in different regions and historical periods, how conflict is constructed and managed, and how identity and efforts to control water systems, through development, technologies, and institutions, shape one another.

The book analyzes responses to the water crisis as efforts to mitigate water insecurity and as expressions of collective identity that legitimate, resist, or seek to transform existing inequalities. The chapters focus on different processes that contribute to freshwater scarcity, including land use decisions, pollution, privatization, damming, climate change, discrimination, water management institutions and technology. Case studies are included from North and South America, Africa, Asia, Europe and New Zealand.

Catherine M. Ashcraft is an Assistant Professor in the Department of Natural Resources and the Environment at the University of New Hampshire, USA.

Tamar Mayer is the Robert R. Churchill Professor of Geosciences and Director of both the Rohatyn Center for Global Affairs and the Program in International and Global Studies at Middlebury College, USA.

Earthscan Studies in Water Resource Management

For more information and to view forthcoming titles in this series, please visit the Routledge website: http://www.routledge.com/books/series/ECWRM/

Recently published titles:

Rules, Norms and NGO Advocacy Strategies
Hydropower Development on the Mekong River
Yumiko Yasuda

The Mekong: A Socio-legal Approach to River Basin Development
Ben Boer, Philip Hirsch, Fleur Johns, Ben Saul and Natalia Scurrah

Trans-jurisdictional Water Law and Governance
Edited by Janice Gray, Cameron Holley and Rosemary Rayfuse

Water and Rural Communities
Local Politics, Meaning and Place
Lia Bryant with Jodie George

Land and Hydropolitics in the Nile River Basin
Challenges and new investments
Edited by Emil Sandström, Anders Jägerskog and Terje Oestigaard

Legal Frameworks for Transparency in Water Utilities Regulation
A comparative perspective
Mohamad Mova Al'Afghani

Water Regimes
Beyond the public and private sector debate
Edited by Dominique Lorrain and Franck Poupeau

Negotiating for Water Resources
Bridging Transboundary River Basins
Andrea Haefner

Environmental Water Markets and Regulation
A comparative legal approach
Katherine Owens

The Politics of Fresh Water
Access, conflict and identity
Edited by Catherine M. Ashcraft and Tamar Mayer

The Politics of Fresh Water

Access, conflict and identity

Edited by Catherine M. Ashcraft and Tamar Mayer

Routledge
Taylor & Francis Group
LONDON AND NEW YORK

earthscan
from Routledge

First published 2017 by Routledge

2 Park Square, Milton Park, Abingdon, Oxfordshire OX14 4RN
52 Vanderbilt AVenue, New York, NY 10017

Routledge is an imprint of the Taylor & Francis Group, an informa business

First issued in paperback 2019

British Library Cataloguing-in-Publication Data
A catalogue record for this book is available from the British Library

Library of Congress Cataloging in Publication Data
Names: Ashcraft, Catherine M., editor. | Mayer, Tamar, 1952- editor.
Title: The politics of fresh water : access, conflict and identity / edited by Catherine M. Ashcraft and Tamar Mayer.
Description: London; New York: Routledge, [2017] | Includes bibliographical references and index.
Identifiers: LCCN 2016027943| ISBN 9781138859227 (hbk : alk. paper) | ISBN 9781315717432 (ebk : alk. paper)
Subjects: LCSH: Water supply–Political aspects. | Water resources development–Political aspects. | Watershed management–Political aspects. | Water security–Political aspects. | Water rights.
Classification: LCC HD1691 .P623 2017 | DDC 333.91--dc23
LC record available at https://lccn.loc.gov/2016027943

ISBN: 978-1-138-85922-7 (hbk)
ISBN: 978-0-367-22772-2 (pbk)

Typeset in Bembo
by HWA Text and Data Management, London

For all the McKibben and Mayer boys (Maximilian and Augustine, and Aviad, Daniel, Ori and Tahl)

Contents

List of figures and tables

Figures

Tables

Notes on contributors

Catherine M. Ashcraft is an Assistant Professor in the Department of Natural Resources and the Environment at the University of New Hampshire (UNH). She received her PhD from the Massachusetts Institute of Technology in Urban and Regional Planning. She is also a Fellow with the UNH Carsey School of Public Policy. Her research and teaching focus on environmental governance, international water agreements, and negotiation theory and practice. Her current research focuses on decision-making around dams in New England, collaborative planning for climate adaptation, institutional arrangements promoting coastal resilience, and adaptive and integrated approaches to flood management in Europe and New England.

Cynthia J. Bannon is an Associate Professor in Classical Studies at Indiana University. She has written two books on Roman law and society, *Gardens and Neighbors: Private Water Rights in Roman Italy* (University of Michigan Press, 2009) and *The Brothers of Romulus: Fraternal Pietas in Roman Law, Literature, and Society* (Princeton University Press, 1997). Her new book project, *Water Communities in the Roman Empire*, continues her research on Roman law and the environment; it investigates how Romans used law to organize the water supply in different physical and cultural environments of the Empire during the first three centuries of the common era (ca. 27 BC to AD 300).

Coleen A. Fox is a Senior Lecturer in the Department of Geography and the Environmental Studies Program at Dartmouth College. She received her PhD degree from the University of Oregon. Her research and teaching focus on international development, environmental politics in transboundary river basins, the role of indigenous knowledge in river restoration, and the political and cultural dimensions of dam removal in New England.

Jessica K. Graybill is an Associate Professor of Geography and Director of the Russian and Eurasian Studies Program at Colgate University. She holds a PhD in Geography from the University of Washington. Researching the human dimensions of environmental change has led her to engage deeply with political ecology, geographies of emotion and cognitive geography.

Understanding the human dimensions of urban, energy and climate transformations guides her research with students in Utica, New York, and globally in resource extraction regions in the Russian North. Jessica works closely with multiple stakeholders to present the kaleidoscope of meanings attributed to and uses of natural and built environments.

David Groenfeldt is Director, Water-Culture Institute, and Adjunct Associate Professor of Anthropology at University of New Mexico. Starting with his PhD research on irrigation development in India, David has focused on water. He worked with the International Water Management Institute (IWMI) in Sri Lanka, and with nongovernmental organizations (NGOs), consulting firms, and multilateral agencies. As director of Santa Fe Watershed Association (New Mexico) he developed an interest in water ethics and in 2010 established the Water-Culture Institute to examine the links between values and water policies. These concepts are detailed in his 2013 book, *Water Ethics: A Values Approach to Solving the Water Crisis* (Routledge).

Leila M. Harris is an Associate Professor in the Institute for Resources, Environment and Sustainability and in the Institute for Gender, Race, Sexuality and Social Justice at the University of British Columbia (UBC). She also serves as Co-Director for UBC's Program on Water Governance. Dr. Harris's work examines social, cultural and political-economic dimensions of environmental issues, with particular focus on equity concerns. Current research focuses on the intersection of environmental issues and inequality, political ecologies of citizenship, and water politics and governance (including focus on marketization, participation and other recent governance shifts, as well as challenges important for the Canadian context).

Pushpa Iyer is an Associate Professor in the Graduate School of Policy and Management at the Middlebury Institute of International Studies. Dr. Iyer is the founding Director of the Center for Conflict Studies and the Editor-in-Chief for the Center's publications. She has over twenty years of experience in the field of conflict studies. Dr. Iyer specializes in identity conflicts, non-state armed groups, civil wars, peace processes, and peacebuilding in post-war societies. Her current research interests are: non-state armed groups, challenges to peacebuilding, and water conflicts. Dr. Iyer is a long-term advocate for the poor and marginalized communities in Gujarat, India.

Marcos López is an Assistant Professor of Sociology at Bowdoin College. His research, based on ethnographic fieldwork he conducted on strawberry production and among farmworkers in Mexico and California, examines the making of agrarian landscapes, ethnic divisions among Mexican farmworkers, and the role labor diasporas play in forming worker protest. Currently, he is writing a book that analyzes the strategies indigenous farmworkers in Mexico's San Quintín Valley use to catalyze labor resistance.

Francis J. Magilligan attended the University of Wisconsin and received a MS in Water Resources Management and a MS and PhD in Geography. He is a Professor in the Geography Department at Dartmouth College and was appointed the Frank J. Reagan '09 Chair in Policy Studies. He is also a Fellow of the Geological Society of America. Besides quantifying the geomorphic responses to floods, his research concentrates on the hydro-ecological impacts of dams and dam removal. He has served on several NSF Panels and is a co-author of a National Research Council report on future research directions in the Geographical Sciences.

Tamar Mayer is the Robert R. Churchill Professor of Geosciences, and Director of both the Rohatyn Center for Global Affairs and the Program in International and Global Studies at Middlebury College. Her research interests focus on the study of nationalism and national identity in the Middle East and in Muslim Western China, and on the relationship among national identity, landscape and memory. She is the editor of *Jerusalem: Idea and Reality* (Routledge, 2008, with S. Mourad), *Gender Ironies of Nationalism: Sexing the Nation* (Routledge, 2000) and *Women and the Israeli Occupation: The Politics of Change* (Routledge, 1994).

Michael Vincent McGinnis is an Associate Professor at the Middlebury Institute of International Studies (MIIS) in Monterey, California. For fifteen years he directed the Ocean and Coastal Policy Center at University of California, Santa Barbara. At MIIS, McGinnis teaches courses in environmental policy, marine science and policy, and sustainable coastal management. He is the author of *Science and Sensibility: Negotiating an Ecology of Place* (University of California Press, 2016), editor of *Bioregionalism* (Routledge, 1999 and 2005) and author of *Marine Governance: The New Zealand Experience* (Victoria University Press, 2012).

T. S. McMillin is the author of *The Meaning of Rivers: Flow & Reflection in American Literature* (American Land & Life Series, University of Iowa Press, 2011) and *Our Preposterous Use of Literature: Emerson & the Nature of Reading* (University of Illinois Press, 2000), along with numerous articles on the American Transcendentalists. He teaches courses on Environmental Studies and American Literature at Oberlin College and is a member of the United Nations Knowledge Network on Harmony with Nature. His latest research project is centered on the Los Angeles River.

Daanish Mustafa is a Reader in Politics and Environment at the Department of Geography, King's College, London. He obtained his PhD from University of Colorado, Boulder. His research interests have been at the intersection of water resources, environmental hazards and development geography. He also maintains a side interest in critical approaches to understanding geographies of violence and terror. While most of Daanish's research has been based in his native Pakistan, he has also published high quality research based in Central Asia, South and Western United States, Central America and Jordan.

Andrés Nápoli is the Executive Director of Fundación Ambiente y Recursos Naturales (FARN, Argentina). Andrés has a JD from the School of Law of National University of Buenos Aires (UBA) and a LLM in Environmental Law from University of the Basque Country. He is a faculty member of the Graduate Program in Metropolitan Environmental Management of the School of Architecture and Urbanism of the UBA. He is co-chair of the board of reviewers of the environmental law journal, *Suplemento de Derecho Ambiental de La Ley*.

Daniel Ryan is Director of Research at Fundación Ambiente y Recursos Naturales (FARN, Argentina) and coordinator of sustainability studies at the Postgraduate School of the Instituto Tecnológico de Buenos Aires (ITBA). Daniel has a PhD in political science from the Department of Government of the University of Texas at Austin, a LLM in Environmental Law from University of London (UK), and a JD from the Law School of the National University of Cordoba (Argentina).

Chris S. Sneddon is a Professor in the Department of Geography and the Environmental Studies Program at Dartmouth College who received his PhD degree from the University of Minnesota. His research and teaching focus on environmental history, water conflicts, and the hydropolitics of the Mekong basin in Southeast Asia. He recently published a book titled *Concrete Revolution: The Bureau of Reclamation, Cold War Geopolitics and Large Dams* (University of Chicago Press, 2015) examining the geopolitical origins of many of world's large dams. His current research focuses on the political and cultural dimensions of dam removal in New England.

Eve Vogel is an Associate Professor of Geography at the University of Massachusetts. Her research focuses on river management, policy, politics and history. She is especially interested in documenting and fostering efforts to manage rivers to sustain ecosystems while also supporting a diversity of human communities and needs.

Maria Alessandra Woolson brings together the worldviews of science and humanities through research and teaching at intersections of cultural, literary and environmental studies in Latin America. Her current work includes interdisciplinary approaches to ecocriticism and field research in Rapa Nui (Easter Island) that employs bio-cultural diversity as a framework for the study of identity and conservation. She also focuses on pedagogical approaches to sustainability education and intercultural studies, developed over her years as faculty at Middlebury College. She teaches at the University of Vermont for the Honors College and College of Arts & Sciences, where she is a sustainability faculty fellow.

Acknowledgements

This book is the outcome of the first international and interdisciplinary annual conference of the Rohatyn Center for Global Affairs at Middlebury College, which both of us, the editors of this volume, organized in 2013. The extreme generosity of Middlebury College and its many departments and programs made it possible to bring to campus outstanding scholars from different parts of the globe. We thank the many who contributed to making the conference a success, in particular the Rohatyn Center for Global Affairs and its staff. Most of the chapters in this volume were presented at the conference, but we solicited a few additional essays in order to cover some topics that were not addressed at the conference. In addition, we thank our student research assistants Toni Cross, Nick Delehanty, Megan English and Isaiah Back-Gaal, as well as Gregory Woolston for making the maps, Jason McKibben for photography assistance, and Abigail Carroll and Mary Bagg for their help copyediting.

We also want to thank our editors at Routledge, Tim Hardwick and Ashley Wright, for their support and patience.

Tamar Mayer and Catherine Ashcraft

1 The politics of fresh water

Setting the stage

Tamar Mayer and Catherine M. Ashcraft

> Water…
> is the cause at times of life or death, or increase or privation,
> nourishes at times and at others does the contrary;
> at times has a tang, at times is without savor,
> sometimes submerging the valleys with great floods.
> In time and with water, everything changes.
>
> (Attributed to Leonardo da Vinci)

Leonardo da Vinci describes fresh water as a paradox: a natural resource and sustainer for which we have no substitute, but also a resource that at once unites and divides communities. Water is not always available where and when it is needed; nevertheless, it invariably contributes to social and national identity and defines humans' interactions with their environment. At times, too much water wreaks havoc on our efforts to control it; at others, water scarcity drives migration, adaptation, and conflict. In fact, fresh water is defined by the discourse of crisis: the crisis of fragmented freshwater systems, the crisis of freshwater supply, and the crisis of pollution (Speth, 2008).

As this book nears publication, the Flint, Michigan, water crisis is still unfolding. Economically disadvantaged and predominantly African American, the city of Flint was denied access to safe water for an 18-month period. In 2014 Flint switched its municipal water supply source from Lake Huron water (treated by the Detroit Water and Sewerage Department) to Flint River water (treated by the Flint Water Treatment Plant). Before switching back to Lake Huron water, Flint's residents and thousands of children, who are particularly vulnerable to lead poisoning, were exposed to toxic lead levels with permanent consequences for their brain development and other health issues. Flint's crisis illustrates three interrelated themes that are central to any discussion of the politics of fresh water and to this book: access, conflict, and identity.

According to the recent report by the Flint Water Advisory Task Force, Flint's crisis demonstrates failures of technology, government decision-making, and conflict management. Flint's poorly maintained Water Treatment Plant was incapable of producing safe drinking water. Flint Public Works personnel followed neither the corrosion-control requirements intended to limit release of lead from aging pipes, nor did they adhere to lead-monitoring rules. By replacing Flint's locally elected representatives with an emergency manager who had sole authority over municipal decisions, including the decision to switch water sources, the state of Michigan denied Flint's citizens access to meaningful participation in water decision-making processes. Other government officials failed to enforce drinking-water regulations, ignored citizen's concerns about the water's color, taste, and odor, neglected to question increasing evidence of lead exposure, and delayed taking action. Conflict between Flint officials and the Detroit Water and Sewerage Department over water-supply contract terms also contributed to the crisis. In addition to these and other failures mentioned in the Task Force's report, Flint's water crisis teaches an important lesson about community mobilization to challenge environmental injustice (Flint Water Advisory Task Force, 2016). Here, the residents of Flint, with support from others from all over the United States, including President Obama, have successfully challenged the government's inadequate response to the problem and its indifference to their concerns.

Two important insights cut across the chapters in this volume. First, the social, physical, and ecological components of water systems are interconnected, forming a hydro-social system. For example, the decision not to treat Flint's water with anticorrosives affected physical systems, including the water treatment and supply infrastructure, as well as ecological systems and human health. The hydro-social concept emerged from the literature on coupled social–ecological systems (e.g., Berkes, Colding, & Folke, 2003), from literature on space, time, place, society/nature, and justice (e.g., Harvey, 1996), from science and technology studies (e.g., Barnes & Alatout, 2012), and from literature on conflict resolution (e.g., Islam & Susskind, 2012). The contributors in this book explicitly critique normative understanding of hydro-social systems and present new insights into the ways access, conflict, and identity are constructed in these systems.

Second, instead of being inevitable, freshwater crisis is a socially constructed experience, a lived phenomenon (De Rycker & Mohd Don, 2013). Speaking about Flint's water crisis, President Obama said, "This was a manmade disaster. … Sometimes it takes a crisis for everybody to focus their attention. … And when we see it, and we understand it, and we feel it, then maybe we start making a connection with each other" (Obama, 2016). Water scarcity is not simply the result of what nature has to offer but always involves power relations and political decisions. The water crisis is not only about who is granted access to safe, clean water (when, where, and why), but also about the extent to which the shrinking of available fresh water influences people's everyday lives at the national and subnational scales. The water crisis also reflects the impact of modernization and neoliberal policies on identity and sense of community. After all, water is the source of livelihood and survival for all people, in every location, at every geographical scale, and the meaning of access to water is inextricably connected to cultural, societal, and political identities. This book's contributors illuminate the dialectic between water and society, reinforcing issues of access, conflict, and identity.

The chapters in this volume provide analyses of the politics of fresh water from multiple perspectives and provide a common stage on which humanists, policy specialists, and social scientists can discuss the complexity of water issues. Together, the contributors offer a deeper and more integrated understanding of the politics of fresh water than would be possible through any single disciplinary approach, scale, time period, or region. They go beyond examining the current state of the freshwater crisis as they analyze the drivers of change in water systems and propose opportunities for the future. Finally, in these chapters the contributors confront problems with the artificial distinction between "nature" and "culture." They suggest that the separation between humans and their natural environment has eroded the identities of those for whom fresh water is a way of life and, at the same time, damaged both nature and culture. Ultimately, this volume challenges us to connect fresh water's diverse meanings for nature, culture, place, economy, and identity and to reconceptualize human experiences *as* nature.

Politics, water, and access

Despite global targets and significant advances in global access to affordable, safe, and clean water, these improvements are textured and uneven (Gerlak & Wilder, 2012). According to 2013 United Nations (UN) figures, more than 750 million people do not have access to clean water, especially those living in rural areas, and almost 2.5 billion do not have access to adequate sanitation (UNESCO, 2013a). Six to eight million people die annually from the consequences of water-related disasters and diseases. Most of these deaths occur in the Global South. Global water use is growing even faster than populations, at the same time as climate change impacts are making freshwater supply even less predictable and reliable (IPCC, 2013). By 2025 two-thirds of the world's population could be living under water-stressed conditions (UNESCO, 2013b).

There are many causes of the freshwater crisis that could be avoided by better planning, by paying attention to the hydro-social system, and by rethinking the connection between "nature" and "culture." The reasons for the crisis include lack of infrastructure to separate wastewater and drinking water; inadequate technology for filtration and disinfection of drinking water; degradation of freshwater systems through human activities, such as pollution and withdrawals; climate change; social practices that determine access based on social hierarchies; privatization systems that exclude the poor; and lack of institutions that ensure equitable access (Brookes & Carey, 2015; Circle of Blue, 2010; Cox, 2015). The chapters assembled here analyze two kind of processes related to access: those by which these physical, social, and ecological factors interact to determine access to water resources and benefits from its use, as well as access to participate in water decision-making; and those by which textured and uneven access patterns, in turn, shape the hydro-social system, including pathways for challenging injustice.

The first part of this volume focuses on institutions that determine who has access to the benefits of water use and who can participate in decisions regarding water. Strengthening institutions is generally recognized as important for improving water access (e.g., Ostrom, 1990). On one hand, institutions are relatively stable rules and practices, which are constructed to prescribe behavior and create expectations for behavior (Keohane, Haas, & Levy, 1993). On the other, they respond to a shifting reality of internal dynamics and changes in their social, political, and economic environments (Young, 1983). Together the cases in these chapters show how access rules are socially embedded, how they create conflict by privileging access for some over others, and how communities can mobilize to enact institutional change.

Using cases from the Roman Empire, Bannon (Chapter 2) analyzes rules defining who has access to water. Private property rights limited access to water in order to secure the supply for the owners of the rights, but exercise of these rights was tied to community norms of civility. In Venafrum, Italy, fountains provided public access to water, but individuals could also buy personal water rights. Illegal diversions were deterred through the placement of boundary stones and through rules ensuring water's arrival in town. In contrast, in the Ebro River irrigation community in what is now Spain, public water was allocated in proportion to land ownership. Irrigators had a right to divert water in proportion to the amount of land they each irrigated and according to a set schedule. In Chapter 11, Mustafa describes a similar rotational schedule for distributing water in *karez* systems in Balochistan (Pakistan) and Nakhchivan (Azerbaijan). In Nakhchivan, where the *karez* is being resurrected, water access is proportional either to the amount of land owned or to the owner's contribution to *karez* rehabilitation. In another case, from New Zealand, McGinnis discusses *kaitiakitanga*, the *Māori* cultural system of values and beliefs, which determines spatial and temporal access to natural resources and stewardship obligations. New Zealand's decision to reorganize its political boundaries according to basin boundaries incorporated *kaitiakitanga*'s ecosystem perspective.

The river basin approach, as applied for example in New Zealand, is often exalted as a natural and objective unit for better water management, which brings together interdependent and diverse interests within the basin to pursue shared objectives (Powell, 1890; Teclaff, 1967). However, as several chapters in this book demonstrate, the decision to adopt a basin scale system of governance is a political decision between contending values and objectives with unequal consequences for water access (Cook, Cohen, & Norman, 2015; McGinnis, 1999). Ashcraft (Chapter 5), for example, analyzes efforts to organize cooperation at the scale of the entire water system in the Danube and Nile Rivers, which included sharp disagreements over access. In the Danube, countries disagreed over the geographic scope of navigation rules permitting commercial access to ports and ultimately differentiated access according to geographic area (main river or delta area) and whether or not the ship was from a country with territory in the basin. Vogel (Chapter 4), by focusing on the complex histories of the Tennessee, Columbia, and Connecticut River basins, shows how organizing cooperation at the basin scale can spread access to a narrow set of benefits, especially hydropower, to some users at the expense of those who depend on access to other river resources, such as fish, and whose ability to participate in decision-making is limited. In contrast, Ryan and Napoli (Chapter 3) describe how a judicial decision, which created a new basin-wide agency for Argentina's Matanza-Riachuelo basin to clean up and control pollution, opened up opportunities for human rights and environmental nongovernmental organizations (NGOs) to participate in making decisions and shape institutions that govern water access.

Investment in built infrastructure is another common approach to improving access to water. However, technocratic approaches often focus on physical elements of the system and exclude other elements, with negative consequences for access to water's diverse social, cultural, and ecological benefits (Gerlak, 2016; McCartney & Dalton, 2015). Magilligan, Sneddon, and Fox (Chapter 6) compare the hydropower and economic benefits from constructing large dams with the large costs for other kinds of system benefits, such as biodiversity. In Graybill's analysis of resource flows in Sakhalin Island (Chapter 7), the Russian federal government, transnational corporations, and international financial institutions engaged in industrial-scale resource development only minimally consider the socioeconomic impacts on local and indigenous subsistence interests who care about freshwater habitat, salmon health, and drinking water. Similarly, in Sri Lanka and Bali (Chapter 9) and Balochistan (Chapter 11), the technocratic approach to irrigation projects physically reduced access to water use for play, bathing, and washing laundry. McMillin (Chapter 15) describes how flood control infrastructure, diversions, and concrete separate the Los Angeles River and limit access to its water system for recreation and habitat for fish and birds. Groenfeldt (Chapter 9), discussing the case of the Cochita Dam in the U.S. state of New Mexico, contrasts actual, negative ecological impacts against the anticipated income, irrigation, and tourism benefits that never materialized. Even when economic benefits are realized from large infrastructure projects, as described in several chapters (9, 10 and 12), these benefits are often not equitably distributed. For example, in Turkey the areas closest to the Atatürk Dam are the least likely to benefit from improved water supply (Chapter 10).

Economic approaches also seek to improve access to water resources and benefits by reorganizing rights and price signals to more efficiently allocate water and satisfy competing demands (Gibbons, 2013). However, Woolson (Chapter 14) and López (Chapter 13) explain how commodification exacerbates water scarcity for those who already have marginal access to water, especially indigenous people (Piper, 2014). In Rapa Nui (Easter Island, a Chilean territory), commercialization of the aquifer by a private operator and prioritization of water consumption by tourists exclude indigenous peoples from land and water use (Chapter 14). Similarly, López (Chapter 13) describes the processes in the San Quintín Valley in Mexico, whereby water access is prioritized for "higher-value" commercial strawberry production over other "lower-value" human and nonhuman benefits. Agricultural firms dominate participation in committees governing the aquifers, and farm workers are left with limited access to water. In contrast to Rapa Nui, inequitable water access in the San Quintín Valley has mobilized a new indigenous political movement to challenge unjust water access and labor conditions. As illustrated by this and other cases in this volume, inequitable access to water is closely connected to conflict, displacement, and loss.

Politics, water, and conflict

Across all cases in this volume, conflict and cooperation are intertwined in water users' clashing worldviews and interests and the institutions, infrastructure, and economic approaches that determine access to water. The cases in this volume analyze both water's destructive potential to exacerbate conflict and its productive potential to foster cooperation (Priscoli & Wolf, 2009).

Because the values people hold are closely linked to their identities, conflicts over values are often particularly intractable and worsen divisions between particular groups of users (Putnam & Wondolleck, 2003). Several chapters in this volume analyze clashes between peoples' worldviews in different parts of the world. McGinnis (Chapter 8) compares commercialization of natural resources in New Zealand through large-scale dairy and hydropower generation with the resulting pollution and decline of the longfin eel, a totem species for the *Māori*. The eel's decline is an indicator of the conflict between the way different users value water systems as either watersheds or pollution-transporting "wastesheds." In the case of the Yavapai tribe's rejection of the Orme Dam in the U.S. state of Arizona (Chapter 9), Groenfeldt highlights the clash between some water users' rational, economic interpretation of nature and others' priceless cultural values. Similarly, the conflict over lining earthen irrigation ditches in the southwestern United States (Chapter 11) contrasts state authorities' view of water seepage as inefficient waste with the *acequia* communities' view of seepage as an important process contributing to the health of a complex agro-social system.

Conflict resolution scholars and practitioners advise against trying to resolve values-based conflicts directly, and recommend focusing instead on practical strategies for managing parties' specific interests (e.g., Forester, 1999). Several contributors in this volume recommend increasing access for diverse interest groups to participate in water decisions (Chapters 9, 11, and 14), for example by creating new spaces that bring diverse interests together to negotiate water policy. In Ebro, Spain, part of the Roman Empire (Chapter 2), a new irrigation community was created to manage conflict between upstream and downstream irrigators living in different political communities. Similarly, Ryan and Napoli (Chapter 3) analyze the effects of creating a new inter-jurisdictional basin forum for negotiating policy disputes in the Matanza-Riachuelo basin in Argentina. Ashcraft (Chapter 5) identifies the creation of multiple basin forums in the Danube as a conflict management strategy through which different environmental interests pursued their own objectives.

However, Bannon, Ashcraft, and Vogel conclude that creating basin spaces and rules for their administration may also exacerbate conflict. In the case of the Ebro River irrigation community, water was allocated proportionally according to how much land a user irrigated, which privileged the wealthy. A powerful elite enforced water rules in ancient Venafrum, not the water users themselves, which undermined community cohesion and strengthened power hierarchies. In the Tennessee and Columbia Rivers (Chapter 4), basin authorities have provided only a narrow range of benefits to specific interest groups at the expense of other basin interests. Until recently, the creation of separate spaces for negotiation of inland navigation and environmental interests in the Danube River (Chapter 5) made it possible for each group to pursue competing objectives, leading to new conflict.

As these cases demonstrate, creating a forum to unite diverse interests from across the water system does not necessarily foster cooperation among diverse interests or empower dissenting interests to participate meaningfully in water decisions. Several other chapters provide examples of institutional mechanisms that have been effective at doing so. A decision by Argentina's Supreme Court (Chapter 3), for example, empowered environmental and human rights NGOs to participate in developing water quality standards. In describing disputes over large dams between economic, environmental, and human rights interests, Magilligan et al. (Chapter 6) identify environmental laws in the United States that mandate public participation as important mechanisms for including interests opposed to large dams. As both Bannon and Ashcraft discuss in their respective chapters, decision-making rules are another way to manage conflict by limiting the power of a majority to act contrary to dissenting interests.

In other cases, delegating conflict resolution to specific individuals was effective for managing conflict. In the Ebro irrigation community (Chapter 2) executive officials were elected from among the irrigators to ensure water use rules were followed and to address violations. Importantly, the irrigators could hold these officials accountable in court if they did not properly execute their authority. In the *karez* and *acequia* cases (Chapter 11) a designated community leader also served an important function in managing conflicts.

Recognizing and integrating differences among water users' interests and priorities is another strategy to effectively manage conflicts (Mnookin, Peppet, & Tulumello, 2000; Susskind, 2014). For example, in the Ebro River (Chapter 2) unequal rights to water were balanced by unequal maintenance obligations. More rights entailed more maintenance responsibility. In the Danube water protection negotiations (Chapter 5) countries integrated their different preferences and priorities for the legal nature of the agreement and the scope of issues on which they would cooperate. In contrast, Ashcraft identifies the absence of strategies to consider tradeoffs as a cause of ineffective conflict management in the Nile. Magilligan et al. (Chapter 6) identify the same problem in the exclusive focus of some large dam investors on greenhouse gas emissions, while neglecting other ecological and social costs.

As these cases demonstrate, when access patterns exacerbate inequities within a community of users, conflict often escalates and divides people and countries. However, water can also bring a community together, for example when people adopt strategies that manage conflict effectively or unite in solidarity to oppose injustice. The next section introduces the dynamics by which politics, water, identity, and community interact.

Politics, water, and identity

Most chapters in this volume underscore the inseparability of water and identity. Water is not only an economic resource to be bought and sold or a necessary element for sustenance of humans, plants, and animals, but it is also a crucial source of identity and community. Whether abundant or scarce, water is the central agent that both shapes the social and physical organization of many communities around the world and reflects the identities, worldviews, and even religious perceptions of their people (Østigård, 2009). This has been the case for nomads, herders, agriculturalists, and pastoralists whose days have been consumed by the search for water and grass. It has also been the case in water-based communities for whom fishing is a way of life, and for those communities in which water is part of the broader cultural landscape (Martin & Trigger, 2015) or the focus of spirituality, mythology, and cosmology. Indeed, many cultures draw their power from water (Boelens, 2014) not simply as a tangible resource or necessary physical substance, but rather as a process from which things occur (Linton, 2010). The focus on water's varying meanings, its tangible and non-tangible qualities, often intensifies as water becomes less available due to commodification and the politics associated with it.

As many chapters in this volume suggest, water participates in the construction of *place* in all corners of the globe and reinforces attachment to both place and community. Water, whether on Sakhalin Island (Chapter 7), in New Zealand (Chapter 8), New Mexico (Chapter 9), Gujarat (Chapter 12), or Rapa Nui (Chapter 14), as Graybill, McGinnis, Groenfeldt, Iyer, and Woolson respectively show, anchors indigenous identities, mythologies, and creation stories. For the *Māori*, Hopi, or the Rapanui, water is the source of spiritual life and defines *place*. At times, the river is the center for spiritual life, where sacred ashrams now stand (Chapter 12), and at others, it is believed to bring together sky and earth to produce a complete whole (Chapter 14). Because place is always the sum of people's experiences and is imbued with values as well as ideology and history, water defines and gives meaning to those who derive their identity from place. Indeed, when water is perceived as sacred, place becomes sacred as well.

But the attachment to place and the identity associated with water-based communities do not have to come from the spiritual powers of water. They can come also from the ecological habitat to which indigenous and subsistence communities are integral. When they depend on the natural environment for survival, as is the case for hunting and fishing communities on Sakhalin (Chapter 7) and the fishing and pastoralist *Māori* communities in New Zealand (Chapter 8), water is the provider and as such shapes culture, identity, and attachment to place. Without water there will be no salmon or longfin eel (*tuna*), nor will there be pastures for sheep and other cattle to graze—all of which are inseparable from identities and cultures that have developed in Russia's Far East and New Zealand, for example. These places are associated in the minds of the locals and outsiders with these gifts of water.

Water can create community and help construct identity in other ways as well. Water can bring people together as they use fresh water for recreation and agriculture (Chapters 11, 12, and 13). In Balochistan and Nakhchivan (Chapter 11) and also in San Quintín Valley in the Mexican state of Baja California (Chapter 13), a sense of community was derived from the availability of water and different activities associated with it. In the Asian examples the *karez* irrigation system provided food security, social capital, and cultural grounding. Because the *karez* also served as a focal point of community gathering and interaction, which was particularly important for women and children, these irrigation systems, Mustafa argues, became a place of social transactions and sites for community building. In the San Quintín Valley, the community's attachment to place developed before water-intensive, market-oriented agriculture took hold in the area. Then, members of the community bathed and played in the spring water that flowed through the valley, interacting with one another and reinforcing their community identity. Not only did these springs quench the thirst of animals and humans but they were also sites for community enjoyment. Rules that allocate responsibility for maintaining community water systems and enforcing water rules also bring people together (Chapters 2 and 11). In antiquity, Bannon writes in Chapter 2, the autonomous Ebro River irrigation community was involved in maintaining the aqueduct system and self-policing water use. As described earlier, creating space for managing a whole water system, such as river basin organization, can also bring together a new community of diverse users who share it.

In other cases water unites a community to oppose water diversion, privatization, damming, and pollution. In New Zealand (Chapter 8) activists are coming together to challenge the dairy industry and restore the eel. In the San Quintín Valley (Chapter 13) indigenous migrant farm workers have successfully used limited access to water to mobilize and demand better labor conditions from the state. In some cases (Chapter 6 and 7), the affected community is joined by a coalition of activists from community organizations and from regional and international NGOs. The anti-dam movement has become a global social movement that connects many local and national communities to form a larger, global community, which can act in support of the local community. In other words, opposition to neoliberal projects that focus on diversion and efficient management of water may unite and create communities at different geographical scales.

Although opposition sometimes brings communities together, many of the chapters in this volume explore the ways by which the massive build-up of large dams in the post-World War II period creates new conflicts and escalates existing conflicts within communities. Using examples from the American Southwest and India, Groenfeldt argues that the introduction of new built technologies is not simply an attack on culture but a real war. When traditional views of water are at odds with the dominant society's beliefs, with very few exceptions, the dominant society almost always prevails. In southeast Turkey, as Harris writes in Chapter 10, the multi-dam project on the Tigris and Euphrates Rivers exacerbates the long simmering Kurdish conflict. The dam's construction has required evacuation from rural areas of the large Kurdish ethnonational minority, which had already been poorly treated in Turkey for decades, and irrigation projects prioritize water access for Arab over Kurdish farmers. The evacuations and ensuing conflicts made the dam and water development project the central node of the Turkish–Kurdish conflict. Similar to Gujarat, India, the unequal distribution of water cannot be separated from gender, religion, and other social differences (Chapter 12). Social hierarchy determines access to and management of water and also determines access to markets, even at the household level, leaving women, ethnic minorities, and the poor more vulnerable to the negative impacts of development. This is the case also in the San Quintín Valley, where the interests of ethnic minorities are moot when those in power exert their will.

Changes in economies and neoliberal commercialization of water systems is also at odds with traditional worldviews and water management practices, leading to much conflict at different geographical scales (Alatout, 2008; Boelens, Getches, & Gil, 2010; Homer-Dixon, 2014; Zwarteveen & Boelens, 2014). The indigenous population in Russia's Far East, for example, whose habitat has been heavily affected by on-and-offshore hydrocarbon explorations (Chapter 7), has struggled to keep alive its traditional way of life, which evolved around salmon. As hydrocarbon explorations will surely continue and intensify, the integrity of both the ecological habitat and its indigenous communities will further erode, leading to displacement and loss. Local and indigenous communities also engage in activities that hurt socio-hydro systems, as seen in Sakhalin. Similarly, even though the cultural heritage of New Zealand's *Māori* (Chapter 8) embraces the connection between people and the ecosystem within which they live, some *Māori* are now involved in commercial enterprises that commodify nature and thus threaten the very rivers on which *Māori* rely for their culture. The conflicts between tradition and modernity may not be reconcilable unless we develop new ways of thinking about rivers and their connected systems.

The concluding chapter in this volume challenges us to rethink what rivers are, how we write about them, and what nature really is. McMillin (Chapter 15) uses the example of the Los Angeles River to argue that all rivers, and as this book shows, all water systems, have an identity, even when their course and character change, even when built infrastructure separates them from the ecological and social systems in which they are involved and these rivers appear to have vanished. All rivers and all water systems, even those that have been manipulated as a result of economic development, have histories, lived experiences, and therefore meanings. Even when all these bodies of fresh water are absent they are actually present. As the Los Angeles River illustrates, there must be other, more helpful ways to think about water systems. If we want to improve them, McMillin suggests we need to better connect water's diverse meanings for nature, society, economy, place, and identity. By focusing on the confluence of these meanings we can revitalize and restore, not simply the rivers, but our thinking about what nature is and "restore thinking to nature (and 'nature' to thinking)" (page 235, in this volume). The false separation between water's meanings is at the root of the *Politics of Fresh Water*.

Works cited

Alatout, S. (2008). "States" of scarcity: water, space, and identity politics in Israel, 1948–59. *Environment and Planning D: Society and Space*, *26*(6), 959–982.

Barnes, J., & Alatout, S. (2012). Water worlds: Introduction to the special issue of *Social Studies of Science*. *Social Studies of Science*, *42*(4), 483–488.

Berkes, F., Colding, J., & Folke, C. (2003). *Navigating social–ecological systems*. Cambridge, UK: Cambridge University Press.

Boelens, R. (2014). Cultural politics and the hydrosocial cycle: Water, power and identity in the Andean highlands. *Geoforum*, *57*, 234–247.

Boelens, R., Getches, D. H., & Gil, J. A. G. (2010). *Out of the mainstream: Water rights, politics and identity*. London: Earthscan.

Brookes, J. D., & Carey, C. C. (2015). Ensure availability and sustainable management of water and sanitation for all. *UN Chronicle*, *51*(4), 15–16.

Circle of Blue. (2010). Water news: Experts name the Top 19 solutions to the global freshwater crisis. Retrieved from http://www.circleofblue.org/2010/world/experts-name-the-top-19-solutions-to-the-global-freshwater-crisis/

Cook, C., Cohen, A., & Norman, E. S. (2015). Conclusion: Negotiating water governance. In E. S. Norman, C. Cook, & A. Cohen (Eds.), *Negotiating water governance: Why the politics of scale matter* (pp. 299–307). Abingdon, UK: Routledge.

Cox, M. (2015). Chapter 10: Comparing water access regimes under conditions of scarcity: The tale of two communities in the United States. In K. Pistor & O. De Schutter (Eds.), *Governing Access to Essential Resources* (pp. 214–234). New York: Columbia University Press.

De Rycker, A., & Mohd Don, Z. (2013). Discourse in crisis, crisis in discourse. In A. De Rycker & Z. Mohd Don (Eds.), *Discourse and crisis: Critical perspectives* (pp.3–65). Amsterdam: John Benjamins Publishing Company.

Flint Water Advisory Task Force. (2016). Flint Water Advisory Task Force—Final Report, Lansing, Michigan.

Forester, J. (1999). Dealing with deep value differences. In L. Susskind, S. McKearnan, & J. Thomas-Larmer (Eds.), *The consensus building handbook: A comprehensive guide to reaching agreement* (pp. 463–493). Thousand Oaks, CA: Sage Publications.

Gerlak, A. K. (2016). Water in international affairs: Heightened attention to equity and rights (review essay). *Global Environmental Politics*, *16*(1), 99–105.

Gerlak, A. K., & Wilder, M. (2012). Exploring the textured landscape of water insecurity and the human right to water. *Environment*, *54*(2), 4–17.

Gibbons, D. C. (2013). *The economic value of water*. Abingdon, UK: Resources for the Future.

Harvey, D. (1996). *Justice, nature and the geography of difference*. Oxford: Blackwell.

Homer-Dixon, T. (2014). Resources and conflict. In C. W. Hughes & Y. M. Lai (Eds.), *Security studies: A reader* (pp. 246–252). New York: Routledge.

IPCC. (2013). Summary for policymakers. In T. F. Stocker, D. Qin, G.-K. Plattner, M. Tignor, S. K. Allen, J. Boschung, A. Nauels, Y. Xia, V. Bex, & P. M. Midgley (Eds.), *Climate change 2013: The physical science basis. Contribution of Working Group I to the Fifth Assessment Report of the Intergovernmental Panel on Climate Change*. Cambridge, UK: Cambridge University Press.

Islam, S., & Susskind, L. (2012). *Water diplomacy: A negotiated approach to managing complex water networks*. Abingdon, UK: RFF Press/Routledge.

Keohane, R. O., Haas, P. M., & Levy, M. A. (1993). The effectiveness of international environmental institutions. In P. M. Haas, R. O. Keohane, & M. A. Levy (Eds.),

Institutions for the Earth: Sources of effective environmental protection (pp. 3–24). Cambridge, MA: MIT Press.

Linton, J. (2010). *What is water?: The history of a modern abstraction*. Vancouver: University of British Columbia Press.

Martin, R. J., & Trigger, D. (2015). "Nothing never change": Mapping land, water and Aboriginal identity in the changing environments of northern Australia's gulf country. *Settler Colonial Studies*, *5*(4), 1–17.

McCartney, M., & Dalton, J. (2015). Built or natural infrastructure—a false dichotomy. Retrieved from https://www.iucn.org/news_homepage/all_news_by_theme/water_news/?19023%2FBuilt-or-natural-infrastructure-a-false-dichotomy

McGinnis, M. V. (1999). Making the watershed connection. *Policy Studies Journal*, *27*(3), 497–501.

Mnookin, R. H., Peppet, S. R., & Tulumello, A. S. (2000). *Beyond winning: Negotiating to create value in deals and disputes*. Cambridge, MA: Belknap Press.

Obama, B. (2016). Remarks by the President to the Flint Community—Flint, MI. Retrieved from https://www.whitehouse.gov/the-press-office/2016/05/04/remarks-president-flint-community-flint-mi

Østigård, T. (2009). *Water, culture and identity: Comparing past and present traditions in the Nile Basin region*. Bergen, Norway: BRIC Press.

Ostrom, E. (1990). *Governing the commons: The evolution of institutions for collective action*. Cambridge, UK: Cambridge University Press.

Piper, K. (2014). *The price of thirst: Global water inequality and the coming chaos*. Minneapolis, MN: University of Minnesota Press.

Powell, J. W. (1890). Institutions for the arid lands. *The Century*, *XL*, 111–116.

Priscoli, J. D., & Wolf, A. T. (2009). *Managing and transforming water conflicts*. Cambridge, UK: Cambridge University Press.

Putnam, L. L., & Wondolleck, J. M. (2003). Intractability: Definitions, dimensions and distinctions. In R. J. Lewicki, B. Gray, & M. Elliott (Eds.), *Making sense of intractable environmental conflicts: Frames and cases* (pp. 35–62). Washington, DC: Island Press.

Speth, J. G. (2008). *The bridge at the edge of the world: Capitalism, the environment, and crossing from crisis to sustainability*. New Haven, CT: Yale University Press.

Susskind, L. (2014). *Good for you, great for me: Finding the trading-zone and winning at win-win negotiation*. Philadelphia, PA: Public Affairs.

Teclaff, L. A. (1967). *The river basin in history and law*. The Hague: Martinus Nijhoff Publishers.

UNESCO. (2013a). Water cooperation facts and figures. Retrieved from http://www.unwater.org/water-cooperation-2013/water-cooperation/facts-and-figures/en/

UNESCO. (2013b). Water factsheets: Water scarcity. Retrieved from http://www.unwater.org/fileadmin/user_upload/watercooperation2013/doc/Factsheets/water_scarcity.pdf

Young, O. R. (1983). Regime dynamics: The rise and fall of international regimes. In S. D. Krasner (Ed.), *International organization* (pp. 93–113). Ithaca, NY: Cornell University Press.

Zwarteveen, M. Z., & Boelens, R. (2014). Defining, researching and struggling for water justice: Some conceptual building blocks for research and action. *Water International*, *39*(2), 143–158.

2 Ancient Roman water rights and commons theory

Cynthia J. Bannon

The Romans are famous for building majestic aqueducts, but the institutional infrastructure, especially law, is as significant as the aqueducts themselves since technology alone was not sufficient to resolve competition over the water supply. In fact, the English word "rival" derives from the Roman word *rivalis*, or "a person who shared a canal." To manage this competition, the Romans developed water rights for both public water systems supplied by aqueducts and for shared water sources on private land. This paper evaluates Roman water rights in three case studies from the early Roman Empire (ca. 27 BC – AD 250): the private property regime of servitudes, a small public aqueduct in Venafrum, Italy, and an irrigation community on the Ebro River in Spain. Each case study analyzes ownership, access, community, and enforcement. The analysis shows that water rights were most socially adequate and most effective at minimizing the problems of free-riders and defection when they were integrated into local communities.

Modern theory/ancient sources

The Roman Empire was neither small nor homogenous: it stretched from Gibraltar to the Euphrates, Mauretania to England, embracing diverse environmental and cultural contexts. The case studies in this paper involve comparable regions (central Italy, northeastern Spain) with similar climates: abundant water during autumn rains and spring snow melts, low water during mostly rainless summers (Al-Mudayna, 1991, p. 16; Horden & Purcell, 2000, p. 14). Culturally, these regions correspond significantly as well, since they adopted Roman cultural practices, including Roman law, early in the history of their contact with Rome.

Water law has a long history in ancient Rome. The earliest evidence appears in the *Twelve Tables*, Rome's first written law code (ca. 450 BC), which probably included property rights (servitudes) to manage water from natural sources on privately owned land (Bannon, 2009, p. 14). In 312 BC, the Romans built the first aqueduct (Aqua Claudia), bringing an open-access public water supply to Rome. Some water from the city aqueducts was delivered to private individuals under special grants administered by city magistrates (Bruun, 2000, pp. 585–587). Water was also illegally diverted from the city aqueducts, despite repeated efforts to curb the practice (e.g., Livy 39.44.4). By the second century BC, there was an additional legal mechanism to regulate the water supply, the praetorian interdicts (implemented by the praetor, one of Rome's magistrates), to protect access to natural sources, first for public rivers then also for private water sources (Alburquerque, 2002, pp. 217–238 and generally). During the reign of Emperor Augustus (27 BC–AD 14), Rome's aqueducts became subject to new laws, as described in Frontinus's treatise, "On the Water Management of the City of Rome" (ca. AD 97–100; Rodgers, 2004, p. 7),[1] which will provide context for the case studies. Throughout the Roman era, Romans used law to manage competition over local water supplies, both natural and artificial.

Romans classified water sources as public or private, although neither category aligns precisely with modern legal notions of public and private (Cole & Ostrom 2012, pp. 41–42). The category of public property included some sources that, in modern terms, would be considered open access, or a commons. The term *res communes* first appears in its modern sense as a commons in the writing of the late Classical jurist Marcian when denoting flowing water, air, sea, seashore (*D.* 1.8.2.1, with De Marco, 2004, pp. 17–18, 23; Zoz, 1999, pp. 89–90). Ordinarily in Roman law, *communis* describes private property owned by several individuals, while common pool resources are owned either by a community (*universitas*) or by no one (*res nullius*) (Zoz, 1999, pp. 35–37; De Marco, 2004, pp. 13–14). Moreover, privately owned water was often treated as if it were a common resource (Bannon, 2009, pp. 47–57).

The incompatibility of modern and ancient categories is unsurprising because the concept of a commons is historically contingent (Cox, 1985), informed by cultural and social practices (Ostrom, 1990). The Roman evidence provides an opportunity to examine a culturally and historically specific application of law to a scarce resource that was, if not always purely open access, usually shared among individuals and groups with conflicting interests. The Roman case is of particular relevance to modern environmental policy because lawyers and economists invoke Roman legal categories as the basis for property rights in the environment. In turn, modern scholarship on common pool resources helps to interpret Roman water rights, supplementing incomplete ancient evidence and guiding evaluation of Roman law's impact.

Sources from the Roman world are less complete than those from later eras. For example, two of the case studies are based on inscriptions, Latin texts written on durable materials for public display and/or archives: a decree on the Venafrum aqueduct and bylaws for the Ebro irrigation community.[2] Neither inscription survives intact—only about half of the Venafrum decree and two thirds of the Ebro law. Sources for the historical contexts are patchy, too: scarce mention of Venafrum occurs in Roman literature, and few inscriptions or architectural remains survive from the site (Capini, 1991). For the Ebro community, evidence is circumstantial, as its location can only be approximated based on toponyms in the inscription and its find spot. Written by and for elites, literary sources carry bias. Legal sources are fragmentary in a different way. The main source of Roman private law is the

Digest of Justinian (AD 530–533), a compilation of excerpts from legal literature of the preceding seven centuries, which scholars use with discretion to reconstruct the law of the early Empire, as the case studies illustrate.

Case study 1: servitudes

Servitudes are the earliest form of water rights in Roman law; four servitudes emerge from the *Twelve Tables*—three for access *(iter, via, actus)* and one for watercourses (*aquaeductus*) (*XII Tab.* 7.8 with Bannon, 2009, pp. 13–15). *Aquaeductus* was a property right attached to private land (the dominant estate), that governed water on land owned by someone else (the servient estate).[3] This one-sided arrangement provided access to water with no fee for use or obligatory exchange. But the legal relationship imposed reciprocal rights and duties on parties. One could only establish a servitude on adjacent property (*D.* 8.3.7.1 Paul.).[4] Informal sharing of water was prohibited (*D.* 8.3.24 Pompon.).[5] A servitude was an indivisible right permanently attached to land and transferred with it in sale and inheritance, but the resource itself was subject to division such that several landowners had access to the same resource, often implemented by schedules (Bannon, 2009, pp. 83–99). These legal rules limited access in a way that conserved water and secured the water supply for both property owners (Bannon, 2009, pp. 48–57).

Romans exercised servitudes within small communities where normative behavior reinforced and defined the legal right to water, as explained by the jurist Celsus:

> If a right of way across anyone's property is granted without qualification or left as legacy to a man, he will be allowed to walk or drive [animals] across any part of it, but respectfully: for certain things are assumed tacitly in the expression. He is not allowed to walk or drive through the villa itself or among the vineyards, when it is equally convenient to do so in another part with less harm to the servient estate. But it is agreed that he should walk and drive where he first directed his path and he should not have power to change it in future. So it seemed right also to Sabinus who used the example of a watercourse which a man had been allowed to route it wherever he liked, after it had been built, he was not allowed to change [it]. It is right that this [rule] be applied to right of way.
>
> (*D.* 8.1.9 Celsus).[6]

The servitude owner must exercise his right respectfully, *civiliter modo*, echoing the Latin term, *ius civile*, the law for Roman citizens, and implying inherent values shared by this community. Celsus explains, "for certain things are assumed tacitly in the term" (*D.* 8.1.9 Celsus, with Bannon, 2009, pp. 103–116). Celsus balances the rights of the two landowners: the owner of the servient estate must grant access to the owner of the dominant estate, but the owner of the dominant estate must exercise his right without damage to the servient estate.

Beyond the two parties, the broader local community also played a role in regulating servitudes (Bannon, 2009, pp. 16–19, 113–117). Repairs to a canal might affect the broader community because the servitude holder could approach his canal by the shortest route across anyone's property. In addition, this community provided witnesses when longstanding use of a water source was recognized as a servitude or when a servitude was challenged under the use-it-or-lose-it rule in which a servitude holder lost his right after failing to exercise it for a prescribed period (two years)—though anyone in the servitude holder's household was free to use it. To prove the rule, the neighbors had to know the servitude holder as well as his slaves and tenants. This rule also assumed informed and self-conscious social practices, as the person using the water had to believe he was exercising a servitude, and he had to know his own legal rights as well.

When these social controls failed, the parties to a servitude had remedies in Roman private law. The servitude holder could claim ownership through *vindicatio* and win compensation for damage to his property caused by interference with his servitude (Bannon, 2009, pp. 159–166; Grosso, 1969, pp. 279–301). In turn, a landowner could contest a servitude or sue for compensation for damage resulting from exercise of a servitude.[7]

Servitudes were fully integrated into both the social and legal practices of Roman landowners. They provided a framework for articulating community norms for shared resources, a framework reinforced by law and in turn shaped by social practice. Servitudes were an effective way of managing a local water supply because they empowered people using the water to design and enforce the legal right to the water (Bannon 2009, pp. 103–116). Servitudes could level the playing field for landowners of modest and lavish means, although Roman private law favored the elite, because servitudes embodied a basic equity among landowners expressed in both reciprocal rights and social values—in particular, the belief that the whole community benefited when all farms were successful (Bannon, 2009, pp. 125–135).

Case study 2: the small aqueduct at Venafrum

Civic aqueducts were fairly common in cities and towns across the Roman Empire. In Venafrum (Venafro) in south-central Italy, the aqueduct brought water from springs at Volturno to the city, as depicted in Figure 2.1 (Galli & Naso, 2009, pp. 142–143; Maiuri, Cimorelli, & Frediani, 1938). The aqueduct and its water were public property because it was built by the emperor Augustus in 16–10 BC, in connection with settling veterans there (Capini, 1991, pp. 29–31; Mommsen, 1850, pp. 287–290, 295–296, 325–326; cf. Zoz, 1999, p. 94). A long, partially preserved inscription, the Venafrum decree documents the administration of the aqueduct's water. Comparison of aqueduct administration at Venafrum with that in Rome reveals typical features of civic administration as well as its limitations.[8]

There were two types of access to water from the Venafrum aqueduct. In town, constantly flowing fountains provided open access.[9] Private individuals could

Figure 2.1 Route of the Augustan aqueduct to Venafrum

Source: Cartography by Gregory Woolston using data from Cees Passchier retrieved from www.ROMAQ.org, and Natural Earth.

arrange rights of access to water for property located both in and outside the town. A pair of mayor-like executive officials called the *duoviri* had authority to:

1 divide and allot the water in order to sell it, *distribuere discribere vendundi causa* (ll. 37–38)
2 impose and establish a fee for use, *ei rei vectigal inponere consti/tuere* (ll. 38–39)
3 make a contract governing these allocations, *legemque ei dicere ...* (ll. 39–43).[10]

Individuals paid a fee for access, and their rights were governed by contract. Although the decree refers to "selling" the water (*vendundi causa*), the contract is not a true sale since individuals did not acquire ownership, only a temporary personal right (that is, not a property right) (Mommsen, 1850, pp. 314–315). In addition to legal rights, rules about the aqueduct's infrastructure limited access. For example, pipes delivering water from the aqueduct could not exceed fifty feet (ll. 43–4), limiting access much like the restriction of servitudes to adjacent property (Cursi, 2007, pp. 124–125). But rules also restricted the flow of water by preventing the Venturi effect, which would increase the volume of flow (Rodgers, 2004, p. 295). Boundary stones along the route identified the aqueduct as public property and were located where people were most likely to attempt puncturing the aqueduct and diverting its water (Mommsen, 1850, p. 293). The location of these stones suggests they were intended to deter free-riders.

The mechanisms for controlling access at Venafrum parallel the administration of Rome's aqueducts, as described in Frontinus's treatise on the aqueducts. The contracts for access in Venafrum match Roman practice during the Republican period, when magistrates (censors, or aediles) arranged temporary personal rights to private delivery for a fee, *ius dandae vendendaeve aquae* (Front. *Aq.* 94). By the time Frontinus was writing, only the emperor authorized individual grants, which his administration then implemented (Front. *Aq.* 99, 105–106). Rome's aqueducts had structural limits similar to those at Venafrum (boundary stones and a fifty-foot limit on pipes) as well as additional safeguards: individual grants specified the size of pipe fittings to control water volume, and private taps were allowed only on reservoirs, not directly on the aqueduct's channel (Front. *Aq.* 105–106, 106.3, 6; Rodgers, 2004, p. 290).

The Venafrum aqueduct served a geographically, socially, and economically diverse community. At its most inclusive, the community included residents of the town and its territory near the aqueduct. All town residents enjoyed open access from continuously running fountains, while some inside and outside town enjoyed individual grants. The decree anticipates conflict between townspeople and those living outside town in several provisions: the placement of boundary stones, the rule on pipes, noted previously, and instructions to owners of private property traversed by the Venafrum aqueduct not to prevent the water from arriving in town (*quominus ea aqua ire fluere ducive possit*, ll.20 & 33–36). Individual grants might also generate conflict between town residents with individual contracts and those without, dividing the community on the basis of economic status (ability to pay the user fee) and political and social connections. The *duoviri* could have used their position to cheat the system rather than enforcing the law, another area of possible corruption.

Comparison with Rome suggests that there was also corruption in the water administration at Venafrum. Frontinus laments depredations on the city aqueducts, laying blame for illegal taps on both individual landowners and the *aquarii*, lower-level officials in imperial administration who created false accounts and sold water off the books when individual grants expired (*Aq*. 74.3, 76.3, 109.2, 114, 128.2). In recounting contemporary corruption, Frontinus praises the old days when Roman administrators privileged the public good over private interests (*Aq*. 95). Even allowing for rhetorical shaping, there was clearly a problem with defectors and free-riders on the city aqueducts, and the problem was not new in Frontinus's era (Front. *Aq*. 76.1, 97.3, with Peachin, 2004, pp. 66–77). The divided community at Venafrum likely experienced similar problems, although part of the aqueduct's route lay underground (*per quem locum subve quo loco specus eius aquae p[erve]nit*, ll. 17–18), where breaking in for purposes of diverting water may have been more difficult, if not impossible. *Duoviri* could have sold water outside the system or manipulated the contracts to their own or to their friends' advantage, much as Roman watermen did.

The Venafrum decree outlines a framework for controlling corruption and enforcing its provisions. The aqueduct's administration had some built-in accountability: it was governed by a law enacted by the *decuriones*, or town councilmen (ll. 41–42). Enacting this law required a two-thirds vote of the town council (ll. 39–40), assuring that a minority did not blatantly co-opt the system. Yet, the *decuriones* belonged to the social and economic elite, and some of the *decuriones* doubtless had individual contracts for use, thereby creating conflicts of interest. At Venafrum, water supply was also embedded in the power hierarchy of the town, so there was more incentive and opportunity for individuals to break the system's rules than for the community to collectively enforce them. The group controlling access and enforcement was not coterminous with the community of water users. The divisions within the water community at Venafrum undermined community cohesion that could support a more equitable sharing of the local water supply. Procedures for enforcing the rules in the Venafrum decree further strengthened the power hierarchy.

The Venafrum decree provides only incomplete information about the system of enforcement because of gaps in the inscription. Local courts in Venafrum may have handled some claims, but the text picks up with instructions for a legal action in Rome in the court of the *praetor peregrinus* (*CIL* X.4842 l. 66, with Cursi, 2007, p. 128). The title *praetor peregrinus* means "*praetor* for foreigners," but this magistrate had broader duties, including jurisdiction over the water supply (Brennan, 2000, pp. 100, 218–219; Serrao, 1954). The *praetor peregrinus* held a preliminary hearing to grant a legal remedy and, following the procedure used in Roman private law, to appoint a board of arbiters (*recuperatores*) who would serve as judges in the trial (ll. 66–70). According to the decree, someone (a private individual? a magistrate?) was delegated by the town to bring legal action on the town's behalf (*is cui ex decreto decurionum ... ne/gotium datum erit, agenti*, ll. 65–66). An official may have been assigned to the water supply, as happened in Rome under Augustus, or these appointments could be ad hoc. Unfortunately, the decree breaks off here, but the rules for enforcement allow some conjecture about their aims and effectiveness. First, the provision for a designated prosecutor suggests recognition of a public interest in the water supply. Alternatively, this provision may be an attempt to deter vexatious litigation through public process. Second, the use of *recuperatores* along with the limited number of witnesses (ten only, ll. 67–68) points to an expedited process and the need to decide quickly because water was a critical issue (Nörr, 2008, pp. 132–133).Third, changing the venue to Rome imposed costs on potential litigants that could have exacerbated the bias towards the wealthy and powerful already institutionalized in the system. Venafrum is 159 kilometers from Rome, nearly 100 miles: Google Maps estimates one hour, forty-six minutes on the A1; by the ancient via Appia, along nearly the same route as the A1, travel on foot or by horse or mule cart would have taken somewhat longer. In addition, holding trials in Rome added another layer of social hierarchy as those with contacts in the capital would have had an advantage in convenience and possibly also in navigating the legal system itself.

Overall, in terms of efficient water sharing, the system at Venafrum was less effective than servitudes because it created a centralized administration separate from the community of users. Because powerful people in the community regulated the Venafrum aqueduct rather than the water users themselves, the system lacked the kind of social cohesion that characterized servitudes and made them an effective mechanism for managing a local water supply.

Case study 3: the Ebro irrigation community

The third case study is also based on an inscription, the *lex rivi Hiberiensis* (*lrH*), also known as the Bronze of Agón, from its find spot in northeastern Spain.[11] The inscription documents an irrigation community that diverted water from the Ebro River into a main canal, the *rivus Hiberiensis*, as the inscription calls it (*lrH* para. 1, I.1). The Roman canal may have followed the same route as one of the modern canals depicted in Figure 2.2. The water was then carried by gravity flow to subsidiary canals serving individual properties.[12] The *lrH* dates to the reign of Hadrian (Beltrán Lloris, 2006, pp. 162–163) and belongs to the same legal and cultural context as the Venafrum decree despite the geographical and chronological distance between them. The *lrH* was enacted to resolve a dispute between upstream and downstream irrigators (Beltrán Lloris, 2006, pp. 186–187), and it established an autonomous irrigation community with its own administration, rules of conduct for irrigators, and legal mechanisms for enforcement.

Because the Ebro was a public river, its water was public property, owned by the Roman state even when diverted onto private property (*D.* 43.12.1.8 Ulp.).[13] Although the *lrH* did not specify the legal status of the canal or its water, the form and language of the decree suggests its public character.[14] Yet the rights to this water resemble servitudes, the private law institution. According to the *lrH*, each

Figure 2.2 Irrigation canals and Roman centuriation along the Ebro River, Spain

Source: Cartography by Gregory Woolston using data from Lloris (2006), Natural Earth, and © OpenStreetMap contributors.

member of the irrigation community had a right to water, *ius aquae* (*lrH* para. 1a, I.8), a vague phrase that can designate either private property rights (servitudes) or temporary personal grants (e.g. at Venafrum).[15] Longstanding use served as the basis for these water rights (Nörr, 2008, pp. 119–120), a legal practice that suggests private ownership, since ownership of public property could not be acquired in this way under Roman law.[16] Servitudes functioned as the model for water rights in a Roman community in Spain at Urso (Capogrossi Colognesi, 1966, p. 82, cf. pp. 52–106), and so likely also served as a model for the *lrH*.

In the Ebro irrigation community, access and ownership were bound together as in a servitude, although, unlike in servitudes, water was allocated in proportion to the land it irrigated.[17] Each landowner diverted water from the canal according to a schedule, on a specific day and at a specific time (*lrH* para. 3c. I.43–46, with Beltrán Lloris, 2006, pp. 176–177).[18] Instructions for annual maintenance of the main canal provide evidence for this schedule. Timed to coincide with the period of the Ebro's lowest flow, the annual maintenance took into account environmental constraints so that all the irrigators had a turn at the water supply, regardless of their location along the canal (Beltrán Lloris, 2006, pp. 176–177; cf. Ostrom, 1990, pp. 72–73; cf. Al-Mudayna, 1991, p. 15 for seasonal flow).

While the *lrH* imposed equality of location, its system of proportional allocation had a more complicated impact. Access based on property ownership privileged the wealthy by replicating the economic hierarchy of the community, but it also set an objective standard that applied to everyone equally. Furthermore, it promoted agricultural productivity, because each property enjoyed the same access to water. Additionally, the legal right gave some protection against coercion by the wealthy. Both the protection and the economic incentive could deter defection by aligning individual interests with the community's common good.

The *lrH* created an irrigation community as a means of settling a dispute among landowners living in the territories of two Roman *municipia*, Cascantum and Caesaraugusta (Beltrán Lloris, 2006, pp. 160–162, 165–166, 178–179, 189–190; Nörr, 2008, pp. 119–120). Because the members belonged to different political communities, they could not simply rely on civic institutions. Instead, the *lrH* established an autonomous community with its own officials and institutions. The community was called a *pagus*, a flexible administrative framework used in various parts of the western Roman Empire (Beltrán Lloris, 2006, pp. 195–197; Tarpin, 2002, pp. 232–233). An unspecified number of *curatores* were in charge of maintenance, and two executive officials, each called *magister pagi*, "master of the *pagus*," presided over assemblies, managed finances, and supervised enforcement of the rules.[19] The *magistri pagi* (plural) may have belonged to the local elite, like the *duoviri* at Venafrum. Unlike the *duoviri*, however, they were probably elected from among the irrigators, who shared collective responsibility for governing the irrigation community (Beltrán Lloris, 2006, p. 190).[20]

The irrigators met in an assembly to arrange maintenance, allocate water, and discuss other matters.[21] Each person's opinion carried weight in proportion to his water rights, *dum proportione quan/tum quique aquae ius habent sententiam dicant* (*lrH* para. 1, I.8). Because land-holding indexed water rights, this rule privileged the wealthy, as we might expect in Rome's hierarchical society. But maintenance duties balanced this privilege, since irrigators' contributions to repairs were also in proportion to water rights.[22] Rules governing repairs also balanced individual and collective interest, as neglect of the main canal threatened everyone's water supply, which the *lrH* acknowledged by describing the negative impact of neglected repairs on downstream irrigators, *quibus aqua in rivo defecerit ... / misve merentur rivos paganico[s? ...] / purgare sarcireque debebit in ...* (*lrH* para. 6 II 22–24, cf. II.22–34). The *lrH* thus defined the irrigation community through a balance of rights and duties that both responded to and attempted to redress social inequality among the irrigators.

The *lrH* anticipated challenges to its provisions by outlining penalties and legal process for enforcement. It prescribed fines for failing to provide funds and labor for repairs to the main canal or for failing to perform other maintenance; it imposed both a fine and remediation on upstream irrigators who blocked the canal.[23] Fines were generally low (25 denarii per violation) and paid to the community (*lrH* para. 1b I.9–15), either reflecting high rates of compliance or enabling irrigators to game the system and pay a fine when the costs of performing maintenance on time exceeded the costs of neglecting other work on their land (Ostrom, 1990, pp. 75–76). On the other hand, the *lrH* set a high fine (250 denarii = HS 1,000) for blocking the canal with debris (*lrH* para. 4 II.7, cf. II.3–11), a more severe penalty to mark truly deviant behavior, probably intentional harm rather than careless or calculated disregard.[24]

The entire community participated in enforcing the rules of the Ebro law. According to the *lrH*, the *magistri pagi* had to give landowners formal notice of their obligations for repairs, and this notice itself may have put pressure on the irrigators to cooperate (*lrH* para. 2a I.16–20, with Beltrán Lloris, 2006, p. 188).[25] In turn, the irrigators could hold the *magistri pagi* accountable in court, imposing fines when they broke the rules or failed to prosecute others' violations (*lrH* para. 11a III.8–14, 23–28; cf. Ostrom, 1990, pp. 73–74, 77–78). The award of half the fines to prosecutors provided an incentive for the irrigators to police themselves (*lrH* para. 11a III.8–14, with Beltrán Lloris, 2006, p. 183). All legal claims arising from the *lrH* were subject to local jurisdiction, which facilitated access to legal remedies and encouraged the irrigators to resolve their own conflicts.[26]

In creating the Ebro irrigation community, the *lrH* probably institutionalized longstanding practice (Beltrán Lloris, 2006, pp. 188, 193–194). Roman law was a ready tool in this Spanish community: already in 89 BC, two towns in the Ebro basin used Roman law to settle a dispute about the construction of an aqueduct (Richardson, 1983). While the *lrH* regulated a public river, it looked to the private law institution of servitudes as a model for individual water rights. The *lrH* represented an equitable and effective system for allocating a scarce water supply because it created an autonomous water community that established its own rules and enforced them—two key elements in successful management of local resources (Ostrom, 1990, pp. 59–61).

Conclusions

The three case studies above show that Romans saw law as an important tool for regulating the water supply. In each, law was used to negotiate conflicts of interest regarding the water supply, and water rights were embedded in the social and economic hierarchy of the community. The three cases differ significantly in the degree of local control over water resources. At Venafrum, city officials managed a public water supply, legal claims were heard in Rome, and the community was divided by conflicts over water: open-access fountains versus individual contracts, city residents versus rural landowners. In both servitudes and the Ebro community, the water users controlled the water system. How well did each system work? Only outcomes of servitudes can be evaluated on the basis of legal disputes reported in the *Digest of Justinian*, and this evidence shows that local control countered the destructive effects of private interest (Bannon, 2009). Unfortunately, direct evidence is lacking for the impact of the Venafrum and Ebro decrees. For Venafrum, comparison with Rome's aqueducts suggests that the decree would not have been adequate to manage the aqueduct. The Ebro community, in contrast, embodied many features of the successful Spanish irrigation communities analyzed by Elinor Ostrom (1990, pp. 69–81), and so is likely to have met the challenge of defectors and free-riders. Moreover, by adapting the principles of servitudes to an open-access resource, the Ebro community shows that water rights could meet the challenge of managing a commons and, in turn, that Roman law offers a dynamic model for modern approaches to water rights.

Notes

1 For Frontinus's treatise, translated with introduction, http://www.uvm.edu/~rrodgers/intro.html.
2 On the discovery and description of the Venafrum decree (*CIL* X. 4842), see Pantoni (1960–1961, pp. 156, 161–169); for the Ebro inscription, see *AE* 1993, 1043 = *HEp* 5 (1995) 911; Beltrán Lloris (2006) pp. 148–156.
3 For the legal status of water on private land, see Fiorentini (2003, pp. 72–75).
4 *D.* 8.3.7.1 Paul., *cf. D.* 8.4.7.1 Paul., and *D.* 8.3.5.1 Ulp., *D.* 8.3.3.1–2 Ulp. (*in vicini villa*), *D.* 8.3.23.1 Paul. (*ad fundum vicinum*), *D.* 39.3.17.4 Paul. (*tuus proximus*).
5 *D.* 8.3.24 Pompon., *with* Grosso (1969, pp. 98–99, 104–108); Capogrossi Colognesi (1966, pp. 158–172).
6 *D.* 8.1.9 Celsus: *Si cui simplicius via per fundum cuiuspiam cedatur vel relinquatur, in infinito, videlicet per quamlibet eius partem, ire agere licebit, civiliter modo: nam quaedam in sermone tacite excipiuntur. non enim per villam ipsam nec per medias vineas ire agere sinendus est, cum id aeque commode per alteram partem facere possit minore servientis fundi detrimento. verum constitit, ut qua primum viam direxisset, ea demum ire agere deberet nec amplius mutandae eius potestatem haberet: sicuti Sabino quoque videbatur, qui argumento rivi utebatur, quem primo qualibet ducere licuisset, posteaquam ductus esset, transferre non liceret: quod et in via servandum esse verum est.*
7 *D.* 8.5.2 Ulp., with Kaser (1971, I.437 & 446), and Grosso (1969, pp. 301–303).
8 The Venafrum decree represents a model for the imperial system in Rome since it predates Augustus' reforms to aqueduct administration in Rome (Cursi, 2007, p. 122).

9 See Mommsen (1850, p. 310) for parallels. Cf. Bruun (1991, pp. 110–114); Hodge (1991, pp. 273–303).

10 Individual access outside town is suggested by the mention of pipes attached to it (*CIL* X.4842 ll. 43–4); for parallels, see Wilson (1999). Distribution in town was usually from reservoirs, not directly from the channel. See Hodge (1991, pp. 273–303).

11 See note 3 above.

12 *lrH* paras. 1a, 3a, 3c (I.1–2, I.27–31, I.42–4) with Beltrán Lloris (2006, pp. 166–170). On Roman irrigation works in the Ebro river basin, see Arenillas & Castillo (2003).

13 Because the Ebro flowed year round (Al-Mudayna, 1991, pp. 15–16), Roman law classified it as a public river: Fiorentini (2003, pp. 68–99); Zoz (1999, p. 110).

14 For the public character of the *lrH*, see Nörr (2008, pp. 116–118).

15 The phrase *ius aquae ducendae* is first attested in Front, *Aq.* 106.1: Capogrossi Colognesi (1966, p. 81, n.150). In the *lrH*, each irrigator's voting rights are in proportion to his right to water, *ius aquae* (*lrH* para. 1, I.8), but nothing in the context helps to define the nature of the right: Capogrossi Colognesi (2014, p. 82).

16 Other phrases suggesting private ownership: genitive of possession, *ei cuius aqua fuerit* (*lrH* para. 12a III.17); possibly a dative of possession, *quibus rivus Hiberien[sis ---] / fuerit (lrH* para. 6 bis, II.25–26). The requirement of notice to the household (*lrH* para. 2a I.16–21) suggests that these water rights could be exercised by other than the right-holder, as in servitudes.

17 Proportional allocation is not expressly prescribed in the *lrH*, but it is inferred from comparative evidence from Roman North Africa (an inscription as interpreted by Shaw, 1982, pp. 72, 82–83) and also from later Spanish irrigation communities: Beltrán Lloris (2006, p. 170).

18 Similar arrangements are known, for example, at Tusculum, near Rome, where water from a stream called the Crabra was "allotted in turn by day and delivery gauges" (Front. *Aq.* 9.5). Similarly, in addition to the North African inscription (note 17), there are two inscriptions from Italy with maps of the watercourse and schedule of deliveries (*CIL* 14.3676 Tivoli; *CIL* VI.1261 Rome?, with Mommsen, 1850, p. 307).

19 *Curatores: lrH* para.2 I.17. *Magistri pagi*: *lrH* para. 1a, I.1–8, 2a I.16–21, para. 3c I.38–46, para. 4 I.47–II.3, para. 6bis, 11a; see Beltrán Lloris (2006, pp. 174–176).

20 Elections are inferred from annual term of office by Beltrán Lloris (2006, p 174).

21 Assemblies: *lrH* para. 3c I.38–46 and 4 I.47–II.1 with Beltrán Lloris (2006, pp. 174–178). The assemblies met at various private locations (e.g. a villa, *lrH* para. 4 I.49), reflecting community-based governance.

22 Contributions of work and money for repairs: *omnes pa/gani pro parte (vacant 4) sua quisque praestare debe/ant* (*lrH* paras. 2b, I.24–26, cf. I.21–33). Each irrigator was also responsible for maintaining diversion dams and bridges on his own property so as not to block the main canal (*lrH* para. 3b I.37–38, cf. 34–38). Collaboration on repairs ensured access all along the canal, according to Beltrán Lloris (2006, p. 173).

23 Fines: *lrH* para. 1b, I.9–15 & 3a I.32–30. Repair and fines: *lrH* para. 2b, II.24–32 with Beltrán Lloris, 2006, pp. 178–179; cf. the same rule applied to the city aqueducts, Front. *Aq.* 129).

24 Compare the high fine (HS 100,000) for damage to the city aqueducts also with the presumption of intentional harm (*dolus*), Front. *Aq.* 129 with Rodgers (2004, p. 328).

25 The community hired private contractors (*publicani*) to collect fines and handle property pledged to guarantee appearance by litigants (*lrH* paras. 8–10 II.43–III.14, with Beltrán Lloris, 2006, 180). A fragmentary passage mentions lower-level civic officials, possibly responsible for the community's records: a freedman, *libertus*, and a scribe(?), *tabellarius*, *lrH* para. 7 II.35; Beltrán Lloris (2006, p. 179).

26 *lrH* para. 10, III.4–6 and para. 14, III.29–37 with Nörr (2008, pp. 124–133). The highest fine may have been set at 250 denarii = HS 1,000 in order to retain local jurisdiction, as this is the typical limit for local jurisdiction in Roman municipal laws Nörr (2008, pp. 125).

Works cited

Al-Mudayna. (1991). *Historia de los regadíos en España (... a.c. –1931)*. Madrid, Spain: Ministerio de Agricultura, Pesca y Alimentacion.

Alburquerque, J. M. (2002). *La Protección o Defensa Del Uso Colectivo de las cosas de dominio público*. Madrid: Dykinson.

Arenillas, M., & Castillo, J. C. (2003). Dams from the Roman era in Spain. Analysis of design forms. In S. Huerta, J. de Herrera, A. E. Benevuto, & F. Dragados (Eds.), *Proceedings of the First International Congress on Construction History*, 20–24 January 2003 (pp. 243–257). Madrid: Instituto Juan de Herrera.

Bannon, C. J. (2009). *Gardens and neighbors: Private water rights in Roman Italy*. Ann Arbor, MI: The University of Michigan Press.

Beltrán Lloris, F. (2006). An irrigation decree from Roman Spain: The *lex rivi Hiberiensis*. *JRS, 96,* 147–197.

Brennan, T. C. (2000). *The praetorship in the Roman Republic*. 2 Vols. Oxford, UK: Oxford University Press.

Bruun, C. (1991). *The water supply of ancient Rome: A study of Roman imperial administration*. Helsinki, Finland: Societas Scientarum Fennica.

Bruun, C. (2000). The Roman world. In O. Wikander (Ed.), *Handbook of ancient water technology* (pp. 575–604). Leiden, Netherlands: Brill.

Capini, S. (1991). Venafrum. In AA.VV. *La romanisation du Samnium aux IIe et Ier siècles av. J.-C.*, (pp. 21–33). Naples, Italy: Bibliothèque de l'Institut Français de Naples.

Capogrossi Colognesi, L. (1966). *Ricerche sulla struttura delle servitù d'acqua in diritto romano*. Milan, Italy: Giuffrè.

Capogrossi Colognesi, L. (2014). "La lex rivi Hiberiensis e gli schemi delle servitù d'acqua in diritto romano." In *Lex Rivi Hiberiensis: Diritto e tecnica in una comunità di irrigazione della Spagna romana*, edited by L. Maganzani and C. Buzzacchi (pp. 75–91). Naples, Italy: Jovene Editore.

Cole, D. H., & Ostrom, E. (2012). The variety of property systems and rights in natural resources. In D. H. Cole & E. Ostrom (Eds.), *Property in land and other resources* (pp. 37–64). Cambridge, MA: Lincoln Institute of Land Policy.

Cox, J. B. (1985). No tragedy on the commons. *Environmental Ethics, 7,* 49–61.

Cursi, M. F. (2007). "*L'edictum Augusti* de aquaeductu Venafrano e l'amministrazione delle acque pubbliche." *Samnium, 80,* 20, 121–132.

De Marco, N. (2004). *I* loci publici *dal I al III secolo. Le identificazioni dottrinali, il ruolo dell'usus, gli strumenti di tutela*. Naples, Italy: Satura.

Fiorentini, M. (2003). *Fiumi e mari nell'esperienza giuridica romana*. Milan, Italy: Giuffrè.

Galli, P. A. C., & Naso, J. A. (2009). Unmasking the 1349 earthquake source (southern Italy): Paleoseismological and archaeoseismological indications from the Aquae Iuliae fault. *Journal of Structural Geology, 31,* 128–149.

Grosso, G. (1969). *Le servitù prediali nel diritto romano*. Turin, Italy: Giappichelli.

Hodge, A. T. (1991). *Roman aqueducts and water supply*. London: Duckworth.

Horden, P., & Purcell, N. (2000). *The corrupting sea: A study of Mediterranean history*. Oxford, UK & Malden, MA: Blackwell.

Kaser, M. (1971). *Das römische Privatrecht* (2nd ed.). Munich: Beck.

Maiuri, A., Cimorelli, V., & Frediani, F. (1938). L'acquedotto Augusteo di Venafrum. Riassunto e rilievi a cura dell'Ente Volturno. *Campania Romana: Studi e Materiali, 1,* 169–185. Naples, Italy: Rispoli.

Mommsen, T. (1850). Römische Urkunden. *ZRG, 15.3,* 287–371. Also published as *Gesammelte Schriften* (3.75–124). 8 vols. Berlin: Weidmann, 1907.

Nörr, D. (2008). Prozessuales (und mehr) in der *lex rivi Hiberiensis*. *ZRG, 125,* 108–188.

Ostrom, E. (1990). *Governing the commons: The evolution of institutions for collective action.* Cambridge, UK: Cambridge University Press.

Pantoni, A. (1960–1961). L'editto Augusteo sull-acquedotto di Venafrum e una sua replica alle fonti del volturno. *RPAA, 33,* 155–171.

Peachin, M. (2004). *Frontinus and the curae of the curator aquarum*. Stuttgart, Germany: Steiner.

Richardson, J. S. (1983). The *tabula Contrebiensis*: Roman water law in Spain in the early first century BC. *JRS, 73,* 33–41.

Rodgers, R. H. (Ed.) (2004). *Frontinus: De aquaeductu urbis Romae*. Cambridge, UK: Cambridge University Press.

Serrao, F. (1954). *La "iurisdictio" del pretore peregrino*. Milan, Italy: Giuffrè.

Shaw, B. D. (1982). Lamasba: An ancient irrigation community. *Antiquités africaines, 18,* 61–103.

Tarpin, M. (2002). Vici *et* pagi *dans l'occident romain*. Rome, Italy: École Française de Rome.

Wilson, A. (1999). Deliveries *extra urbem*: Aqueducts and the countryside. *JRA, 12,* 314–331.

Zoz, M. G. (1999). *Riflessioni in tema di* res publicae. Turin, Italy: G. Giappichelli Editore.

3 Legal mobilization and the politics of water pollution

The case of the Matanza-Riachuelo basin in Argentina

Daniel Ryan and Andrés Nápoli

Introduction

This paper explores the impacts of judicialization on environmental politics. Specifically, it analyzes whether and how the Supreme Court's involvement affected the politics of pollution control in the Matanza-Riachuelo basin in Argentina (hereafter, the Riachuelo basin). A conventional legal perspective tends to assume that judicial resolutions can change a state's practices and policies and ensure the realization of rights, including socio-environmental rights. However, a large body of literature stresses the court's limitations when it comes to significant policy and social changes (see, for instance, the influential works of Horowitz, 1977, and Rosenberg, 1991). This paper explores this issue in the context of the Riachuelo basin.

The Matanza-Riachuelo River is considered the most polluted watercourse in Argentina. The basin is located in the southern part of the metropolitan region of Buenos Aires, a densely populated region and one of the most heavily industrialized in Argentina. The federal government, the city of Buenos Aires, the province of Buenos Aires, and fourteen local municipalities hold legal jurisdictions over different parts of the basin, depending on the issue. This creates a very complex governance framework. In 2008, after years of fragmented local protests and ineffective government attempts to clean up the basin, Argentina's Supreme Court ordered all of the governments involved to control environmental pollution, clean up the basin, and provide environmental information to the public.

The court's landmark decision raised high expectations among the Argentine public, especially among the environmental and public interest law communities. The Riachuelo carries strong symbolic value in Argentina as a paradigmatic example of governments' inaction or inability to address socio-environmental degradation and to protect the public's right to a clean and healthy environment. Nevertheless, government compliance with the resolution has been uneven. The governments are implementing a clean-up plan in the basin but have only partially complied with the court's mandates. In fact, the original timelines set by the court have had to be extended, and it is not clear to what extent water quality and pollution levels in the Riachuelo have improved.

We share most of these assessments about the delays and weak government implementation of many aspects of the court's resolution (see, for instance, Nápoli 2013a and 2013b). Despite these serious problems and shortcomings, we still argue that the court's involvement has substantially changed the politics of pollution control in the Riachuelo basin. First, the court's resolution overcame the political inertia that affected the policy process and debate about the Riachuelo's socio-environmental problems. It set the basis for a policy agenda and prompted the governments involved to create a new governance framework for the basin. Building on Sabel and Simon's destabilization argument (2004), we claim that the Supreme Court's involvement ultimately unsettled and opened up an entrenched institutional setting that had systematically failed to address the environmental problems affecting the Riachuelo basin. Second, the judicialization of the dispute changed the structure of opportunities for social advocacy related to the Riachuelo. Specifically, it has strengthened the political leverage of those social actors involved in the judicial procedures, namely professionalized environmental and human rights nongovernmental organizations (NGOs). These groups can now access and affect the policy agenda and process related to the Riachuelo basin in much higher ways than before the dispute's judicialization.

This paper opens with a brief review of the literature on judicial impacts, then concisely describes the Riachuelo case followed by an analysis of the two main impacts of the process of judicialization mentioned above. Finally, it offers comments on potential broader theoretical implications of this analysis.

Conceptualizing judicial impacts

The literature on courts and social change offers different conceptualizations of judicial impacts. Scholars commonly draw a distinction between "direct" and "indirect" effects of judicial involvement (Garavito & Franco, 2010; Gauri & Brinks, 2008; McCann, 1994; Rosenberg, 1991). Direct effects are generally defined as the behavioral and policy changes intended and achieved by a court's rulings. For instance, a judge orders the government to stop a mining project because the environmental impact assessment is deficient. The direct impact of judicialization in this case depends on the level of compliance with the court's resolution. By contrast, indirect effects refer to the consequences of courts' decisions and involvement that are more diffuse and indeterminate in nature. These results affect actors and relationships, usually over time, which were not directly the target of the courts' actions, for instance changes in public opinion about an issue, social mobilization around a particular cause, and government actions to promote change in existing legislation.

The distinction between direct and indirect effects reflects deeper differences in the literature about the role of courts and judicial procedures and, more generally, about how to address and study the consequences of judicialization. Scholars focusing on the direct effects of judicial rulings tend to develop court-centered analysis. By carrying out "impact studies" of court rulings, this literature compares court mandates and promises with actual policy changes resulting from these mandates and promises.[1] On the other hand, scholars focusing on courts' indirect effects tend to develop dispute-centered analysis. In this perspective, the focus is on how legal mobilization and court rulings (including lost cases) shape the political and institutional landscape in which the different actors involved in a policy dispute negotiate with each other.[2] Our paper draws on both approaches. It analyses the impact of the Supreme Court's involvement —a court-centered approach. However, our analysis does not focus on the level of judicial compliance but on how the court's involvement has affected the institutional and political landscape in which the dispute and debate over pollution control of the Riachuelo basin has evolved.

The Matanza-Riachuelo basin case: a brief description

As mentioned above, the Matanza-Riachuelo basin is considered one of the most polluted urban areas in Argentina and Latin America. In fact, according to the Blacksmith Institute, it is one of the thirty most polluted sites in the world.[3] At 75 kilometers long, the Riachuelo River boasts a watershed of 2,240 square kilometers, its basin encompassing part of the city of Buenos Aires, the capital city of Argentina, and fourteen municipal districts in the province of Buenos Aires. Figure 3.1 includes a map of the basin.

The Riachuelo basin is a densely populated region. According to the 2010 national census, approximately 8,212,000 people live in the basin territory, representing around 20 percent of Argentina's total population.[4] A high percentage of the basin's inhabitants, especially those living on the river banks, is poor, lacking access to sewage and water services and living in unsafe slums. Based on data from that year, 10.9 percent of the basin population has unsatisfied basic needs (UBN).[5] Additionally, the basin contains some of the most heavily industrialized areas in the country. It is home to approximately 11,000 industries, many of which have high environmental impact such as those in the Dock Sur Petrochemical District. Moreover, many industries do not use proper treatment procedures or have proper facilities and often dump their effluents directly into the basin's watercourses without treatment. To compound the problems, the basin territory hosts over 300 illegal open landfills in which residential and industrial waste, and sometimes even hazardous waste, is disposed without proper preventive measures or treatment. In short, industrial effluents, untreated sewage discharges, and illegal open landfills are the main sources of pollution affecting the Riachuelo basin and the people living in this region.[6]

Since the 1990s, several governmental initiatives have aimed to clean up the basin and tackle various environmental problems affecting the region. In 1993, the national government, under the administration of President Menem, created an executive committee (*Comité Ejecutor Matanza Riachuelo*) to carry out an environmental management plan for the basin. The secretary of environment at the time, María Julia Alsogaray, famously announced that the Riachuelo would be clean in 1,000 days. By September 1995, the 1,000 days had passed and the plan had not advanced. The Menem Administration then created a new executive committee

Figure 3.1 The Matanza-Riachuelo Basin

Source: Cartography by Gregory Woolston using data from ACUMAR, retrieved from http://www.acumar.gob.ar/mapa-cuenca/ (accessed May 5, 2015), and Natural Earth.

(*Comité Ejecutor del Plan de Gestión Ambiental y de Manejo de la Cuenca Hídrica Matanza Riachuelo*, or CERM), which this time included the city of Buenos Aires and the government of the province of Buenos Aires. However, while CERM functioned as an implementation unit of the environmental management program, it had no regulatory powers and no ability to coordinate the jurisdictions involved.

In 1998, the Inter-American Development Bank (IADB) approved a US$250 million loan to Argentina to partially fund implementation of the basin's environmental management plan, but only a very small part of the loan's funds were used by the successive Argentine governments (Auditoria General de La Nación, 2006). The government carried out some tasks such as cleaning the surface of the river (removal of old ships, etc.) and installing flood control works, but it made no major advances in pollution control or significant improvements in the overall environmental quality of the basin.[7]

During these years, the federal government as well as the governments of the province of Buenos Aires and city of Buenos Aires passed general environmental legislation. Although this legislation did not specifically target the Riachuelo's environmental problems, it did provide legal tools with which to control the pollution affecting the basin. However, experts, activists, and local residents generally agree that enforcement of these pollution control regulations in the basin remained extremely low (Fundación Ciudad, 2002; Defensor del Pueblo de La Nación, 2003). Many have observed that the fragmented nature of state power in the basin presented a major obstacle to policy efforts. Various agencies belonging to the national government, the governments of the city of Buenos Aires and the province of Buenos Aires, and the local governments of the fourteen municipalities hold legal jurisdiction over diverse issues and parts of the territory of the basin without sufficient inter- and intra-jurisdictional coordination (Defensor del Pueblo de La Nación, 2003).

While social awareness and concerns about the Riachuelo's environmental problems had been growing over the years, collective action had been very fragmented. Social protests and mobilizations tended to be reactive, focalizing after pollution or flooding events affected specific areas of the basin. Moreover, after these emergencies had passed, broad popular attention and mobilization usually languished, leaving only a small number of local residents or associations working on these issues (Lanzetta & Spósito, 2003).[8] As collective actions and demands tended to be territorially focused, addressing specific environmental emergencies affecting specific neighborhoods, local associations and neighborhood groups that mobilized around these issues in different parts of the basin had showed low levels of articulation and cooperation (Ryan, 2004). Arguably, the lack of a coordinated and sustained collective demand in response to the socio-environmental problems affecting the Riachuelo greatly limited the political leverage of environmental NGOs and local groups demanding policy changes at the basin level.

By the end of 2002, the National Ombudsman's office formed a working group with a few well known NGOs and other institutions to try to "move up" the issue of the Riachuelo on the national political agenda.[9] This decision was triggered by various complaints received during that year about specific pollution problems affecting basin residents and communities (Ryan, 2004). In August 2003, based on a report produced by the working group, the National Ombudsman issued a resolution declaring the Matanza-Riachuelo basin to be in a state of health and environmental emergency, and requested different agencies and ministers of the national government to take measures to address the problem, including creating a specific inter-jurisdictional agency for cleaning up the basin. The report focused considerable media attention on the Riachuelo, and many high-level government officials publically acknowledged the problems affecting the basin. However, the policy response from the national government, at the time under the presidency of Nestor Kirchner, was poor, and no significant policy measures were taken (Defensor del Pueblo de La Nación, 2005).

In 2004, a group of seventeen residents from *Villa Inflamable* (a highly polluted slum in the Dock Sur at the mouth of the Matanza-Riachuelo) along with health workers from a nearby public hospital filed a legal complaint for environmental damages to the Supreme Court of Justice against forty-four large polluting basin businesses, the national government, and the governments of the province and the city of Buenos Aires. In June 2006, the court declared itself competent to hear the claim for collective environmental damage,[10] and it issued an initial resolution ordering the national government and the governments of the province and the city of Buenos Aires to submit a plan to clean up and to control the environmental pollution in the basin according to National Environmental Law 25.675 (*Mendoza, Beatriz y otros c/ Estado Nacional y otros,* 2006). Furthermore, the court requested the forty-four large polluting businesses to provide information about the environmental impact of their activities and whether they had fulfilled other requirements of existing environmental regulations. A few weeks later, the court granted intervention in the case to the National Ombudsman and several NGOs as interested third parties (*terceros interesados*), who in turn became very active actors in the judicial procedure.

The court also began a process of hearings to gather information about the basin situation and monitor the parties' compliance with its initial resolution. The hearings helped to keep media, government, and public attention on the issue (Nápoli, 2009a). In July 2008, the court issued its final resolution (*Mendoza, Beatriz y otros c/ Estado Nacional y otros,* 2008), a landmark decision ordering the national government and the governments of the city of Buenos Aires and the province of Buenos Aires to develop a program to restore the basin's environment, prevent future environmental damages, and improve quality of life for the population. While the court's resolution gives discretion to the governments involved in terms of program content and how to achieve these objectives, the resolution does list nine key components, or issues, which should be addressed by the program, including the creation of a system of public information, a system of quarterly water quality reports for the basin, expansion of the sewage and potable water system, and monitoring of industrial pollution.

More than five years have passed since the court's 2008 landmark decision, and, as mentioned in the introduction, compliance with the court's resolution has been uneven. The governments involved are finally implementing a clean-up plan in the basin but have only partially complied with the court's objectives and mandates (Nápoli, 2013a, 2013b). Despite these serious shortcomings, the judicializations of the issue and the court's involvement have substantially changed the politics of pollution control in the Riachuelo. In the next sections we develop this argument.

Judicial impacts on the politics of the Riachuelo basin

Overcoming political inertia

The Supreme Court's involvement triggered a new political dynamic around the socio-environmental problems affecting the Riachuelo basin, and this involvement overcame the political inertia that had characterized the issue for decades. A clear example of this change is the creation of the *Autoridad de la*

Cuenca Matanza Riachuelo (ACUMAR), an inter-jurisdictional agency mandated to clean up and control pollution in the Riachuelo basin.

Many experts and activists have stressed that the fragmentation of state authority over the basin has been a key factor in explaining weak pollution control and, more generally, the lack of a proactive agenda for addressing the socio-environmental problems affecting this area (Defensor del Pueblo, 2003; Nápoli, 2009a). Regulatory and enforcement powers over pollution control, water quality, land use regulations in the basin territory (just to mention a few key policy fields) are scattered between agencies belonging to different levels of government (the federal government, the government of the city of Buenos Aires, the government of the province of Buenos Aires, and fourteen municipal governments). In this complex institutional scenario, disputes over jurisdictional issues are extremely common, making the development and implementation of policies and programs at the basin level more difficult (Defensor del Pueblo de La Nación, 2003).

In this context, experts, academics, and environmental activists have been continually demanding inter-jurisdictional coordination of policymaking and policy enforcement in the Riachuelo basin. In fact, in the 2003 Riachuelo report prepared by the National Ombudsman's Working Group, inter-jurisdictional coordination was one of the main claims. However, despite repeated calls for stronger coordination in the basin, no significant advances were made until the Supreme Court's resolution. Arguably, the creation of some sort of an inter-jurisdictional arrangement for controlling pollution in the basin raised different short-term costs for each of the governments involved (giving up enforcement powers, re-allocating funds, etc.), while the political benefits of potential environmental improvements were rather diffuse and more likely to occur in the middle or long term. In this context, no government had strong incentives to promote policy coordination at the basin level. It is not surprising, then, that the political process around the socio-environmental problem affecting the Riachuelo basin was characterized by its inertia, modified only by focalized policy responses addressing specific environmental emergencies.

The court's involvement dramatically changed this dynamic. It placed the socio-environmental problems affecting the Riachuelo basin on the national political agenda. Moreover, it compelled the governing coalitions (as well as the main opposition parties) at the federal government and at the governments of the city and province of Buenos Aires to take a political position in relation to the judicial mandate to clean up the basin.[11] In this way, the court's rulings increased the political costs of maintaining the status quo for the main political actors.

Accordingly, only a few weeks after the court issued its first resolution on the Riachuelo (June 2006), the federal government submitted a bill to Congress creating the ACUMAR. Unlike previous programs and agencies created to clean up the Riachuelo, the ACUMAR was granted powers to regulate, monitor, and intervene in prevention, clean-up, environmental recomposition, and rational management of the basin. Moreover, the ACUMAR's governing board was to consist of representatives from the three main governments involved in the basin: four from the national government, two from the government of the city of

Buenos Aires and two from the government of the province of Buenos Aires. The bill was promptly considered by the National Congress, and, by November 2006, both legislative chambers had already approved it (law 26.168) with the support of the main opposition parties.[12] A month later, in December 2006, the Legislature of the city of Buenos Aires approved the creation of ACUMAR (city law 2.217), and, by February 2007, the Legislature of the province of Buenos Aires had done the same (provincial law 13.168). In both cases, local governing coalitions and main opposition parties supported the laws ratifying the creation of ACUMAR.

In sum, in ten short months, the legislatures and the executives of the three governments created an inter-jurisdictional agency to clean up and control pollution in the Riachuelo basin after decades of policy and legislative inaction on the issue. Furthermore, the laws gained approval with broad political support from the governing coalitions and the main opposition parties, strongly suggesting that prior political inaction was not due to substantive policy differences between the main political parties or coalitions involved.

Several speeches during the legislative debates stressed the key role played by the Supreme Court in this policymaking process. Senator Cristina Fernandez de Kirchner, the wife of President Kirchner and one of the leaders of government legislative block in the Senate stated, "It is worth pointing out that this bill submitted by the national executive has its origin on a request made by the Supreme Court of Justice to the national government and to the rest of the governments with authority and jurisdiction in the Matanza-Riachuelo Basin..." (Cámara de Senadores de la Nación Argentina, October 11, 2006).[13] Legislators from opposition parties made similar statements: "How does this issue [the problems of the Riachuelo], which has existed for many years, become part of the agenda? Because of a terrifying report prepared by the National Ombudsman office and a request made by the Supreme Court of Justice in the Mendoza case..." (Deputy Marta Maffei, ARI, Cámara de Diputados de la Nación, November 15, 2006). As these speeches underscore, the sudden legislative activism in relation to the Riachuelo basin was clearly prompted by the Supreme Court's involvement.

Obviously, the mere creation of ACUMAR did not solve the socio-environmental problems affecting the Riachuelo basin, but it did change the institutional scenario in which the policymaking process surrounding the Riachuelo takes place. Currently, ACUMAR—on behalf of the governments involved in the lawsuit—carries responsibility for developing and implementing measures to control pollution and clean up the basin in order to comply with the court's resolution. Previously, there had been no institutional venue for discussing basin policies and no government authority to be held responsible for the basin's environmental health, but with the creation of ACUMAR, a locus emerged for the basin policymaking process and an agent accountable for the basin's environmental agenda.

Strengthening NGOs' access to the policy process

As mentioned above, collective action and social demands to improve the Riachuelo's environmental quality were historically marked by a high level of fragmentation and discontinuity. Protests and mobilizations tended to be reactive and focalized, following on the heels of environmental emergencies affecting specific areas of the basin. A few local groups and networks as well as some more professionalized environmental NGOs followed the policy agenda in a more systematic way, demanding basin-wide policy changes. However, lack of a broader and sustained social demand to clean up the Riachuelo and prevent pollution greatly limited these groups' political leverage when it came to influencing the policy agenda and policymaking processes.

In this context, the judicialization of the Riachuelo dispute changed the structure of opportunity for social advocacy related to this issue. Supreme Court resolutions have distinctly increased the political leverage of those social actors that are party to the judicial procedures, giving them greater access to and voice in the policy process about the basin's clean-up. The Cuerpo Colegiado is a good example of this. In June 2008, the court created a commission called Cuerpo Colegiado to monitor compliance with the court-established objectives, goals, and deadlines for the clean-up.[14] This monitoring commission was formed by the National Ombudsman office and five NGOs granted legal intervention in the judicial procedure: Fundación Ambiente y Recursos Naturales (FARN), Greenpeace Argentina, Centro de Estudios Legales y Sociales (CELS), Asociación de Vecinos La Boca, and the Asociación Ciudadana por los Derechos Humanos (ACDH).

The Cuerpo Colegiado has no enforcement or regulatory powers, but it can submit reports to the court regarding actions and measures taken by ACUMAR to comply with the court's resolution. In this way, the Cuerpo Colegiado has played a key role in placing issues on the Riachuelo policy agenda. For instance, the Cuerpo Colegiado's advocacy work drove ACUMAR to adopt certain procedures for the relocation of the population at risk. The clean-up of the Riachuelo requires relocating inhabitants of the slums lining the river banks. Approximately 450 families had been already relocated, and about 2,500 more are expected to be relocated in the short term (Nápoli, 2013b). Relocation, however, is an extremely complex process usually fraught with conflict between the local communities and the involved governments. In April 2012, the National Ombudsman, on behalf of the Cuerpo Colegiado, proposed to the Supreme Court a set of guidelines for governing the processes of relocation (Defensor del Pueblo, 2012). The guidelines require ACUMAR and other involved government agencies to establish certain special mechanisms for taking into account the concerns and interests of disadvantaged communities set to be relocated. In December 2012, the court issued a resolution listing the issues that ACUMAR should prioritize in order to foster the basin clean-up, and among those priorities was local residents' right to participate in relocation plans (Corte Suprema, 2012).

Similarly, the Cuerpo Colegiado has taken a central role in developing policy governing the Riachuelo's water discharge system. In 2007, ACUMAR issued a resolution (Resolution 1/2007) establishing limits for liquid effluent discharges. This was later complemented by ACUMAR's resolution 3/2009, which defines water uses and objectives for the Riachuelo basin. Both resolutions, however, received strong criticism from the Cuerpo Colegiado because of their low standards (Nápoli, 2009b).

Many argue that even if basin industries were to fully comply with water quality standards established by ACUMAR, the Riachuelo would still be highly polluted (Estrada Oyuela *et al.*, 2012; Nápoli, 2013). In its December 2012 resolution, the Supreme Court cited the Cuerpo Colegiado's repeated observations related to existing water standards and requested ACUMAR to modify these standards.

As these two examples illustrate, the Cuerpo Colegiado's work substantially increased member NGOs' access to the policy process and their ability to influence the Riachuelo's policy agenda. Given the institutional framework created by the court, government officials in charge of implementing the judicial resolution could not simply ignore these groups as stakeholders in the policy process. In comparison with their situation prior to the judicialization of the dispute, the court's resolutions clearly increased the political leverage of those social actors participating in the judicial procedures.

Most of the NGOs comprising the Cuerpo Colegiado (FARN, Greenpeace, and CELS), however, are not grassroots associations but professionalized, nationally oriented organizations with specialized knowledge and technical skills related to environmental and human rights policy and legal issues.[15] Arguably, most local networks and neighbor associations lack the human and technical resources needed to actively participate in the judicial and monitoring processes of the basin clean-up plans. In this context, then, although the Supreme Court's resolutions have expanded the structure of opportunity for social participation and advocacy in the policymaking process around the Riachuelo, certain types of social actors, namely professionalized NGOs, have been better positioned to take advantage of the opportunities generated by the dispute's judicialization.

Final comments

This paper shows how Supreme Court involvement has affected the politics of water pollution in the Riachuelo basin. First, it reveals how courts can overcome inertia in the political process, placing issues on the policy agenda that political actors are not otherwise addressing. In this regard, our analysis of the Riachuelo case clearly resonates with the literature arguing that courts and litigation can have destabilization effects (Sabel & Simon, 2004; Dixon, 2007). According to this view, judicialization can effectively open up public institutions that have chronically failed to meet their obligations and that have been relatively immune to the conventional political mechanisms of correction (Sabel & Simon, 2004). In the case of environmental problems affecting the Riachuelo basin, the political system was unable or unwilling to address the issue and to unsettle the existing failing institutional framework. Moreover, the social demand for more effective pollution control and better environmental quality in the basin was fragmented and discontinuous, and it did not affect democratic politics in a way that could trigger the needed policy changes. In this context, the Supreme Court's involvement functioned as an exogenous factor that shook the status quo and prompted a process of institutional change in the Riachuelo basin.

Second, our analysis shows that judicialization can play an important role in expanding the structure of opportunity for social advocacy. In the case of the Riachuelo, Supreme Court resolutions have distinctly increased the political leverage of those NGOs that are party to the judicial procedure. Through the creation of the Cuerpo Colegiado, these NGOs and the Ombudsman office have gained much greater access to and—and a more powerful voice in—the policymaking process around the Riachuelo than they enjoyed prior to the dispute's judicialization. This, of course, raises a series of interesting empirical and normative questions about the representativeness and the accountability of the specific associations and groups that gain access to a particular policy process as the result of a dispute's judicialization. In other words, to what extent can these groups or NGOs be considered representative of the broader, diffuse constituencies affected by the policy in question? To whom are they accountable, for both the positive and negative policy consequences that may result from their decisions in judicial procedures? Clearly, these issues deserve further research and analysis.

Notes

1 Rosenberg's well known study (1991) of the U.S. Supreme Court decision on *Brown v. Board of Education of Topeka* is a paradigmatic example of this approach.
2 For a more detailed analysis and comparison between these two approaches, read the exchanges between Rosenberg (1996) and McCann (1996) published by the *Journal of Law and Social Inquiry*.
3 See http://www.blacksmithinstitute.org/wwpp2007/finalReport2007.pdf, cited by Nápoli and García Espil (2010, 177).
4 Data available at the ACUMAR website: http://www.acumar.gov.ar/indicadores/1173/poblacion-segun-sexo-y-edad (accessed May 2, 2015).
5 According to the National Institute of Statistics and Censuses (INDEC), households with unsatisfied basic needs (NBI) are households having at least one of the following indicators of deprivation: overcrowding, disadvantaged housing, poor sanitary condition (lack of toilet), school-age children do not attend school, or subsistence-bounded capacity. Data available at http://www.acumar.gov.ar/indicadores/1203/ (accessed April 29, 2015).
6 For a detailed overview of the geography, history, social economic features, and main environmental problems of the Matanza-Riachuelo basin, see the report prepared by Fundación Ciudad (2002); also, see Nápoli (2009a).
7 A sum of only $7.7 million (out of $250 million, which constituted the total loan) was executed in Riachuelo-related works. During the 2002 Argentine economic crisis, the IADB and the Argentine government (at this time under the administration of President Duhalde), agreed to use part of the loan's remaining funds to cover urgent social programs.
8 Such was the case, for instance, of a toxic "cloud" that affected several elementary schools in the area of the Dock Sur during October–November 2001. This event triggered strong protests among local residents, including street blockades. This incident and the social protest that followed it led to the creation of an environmental control committee in the Dock Sur, with the participation of businesses, governments, and local NGOs. The committed was active during 2002, but by 2003 it had lapsed. Some local associations continued their advocacy work on the issue, but the level of social attention and mobilization dropped significantly. For a more detailed description and analysis of this specific case, see Lanzetta and Spósito (2003) and Ryan (2004).

9 The working group was formed by several universities, NGOs (FARN, Greenpeace, CELS, *Fundación Ciudad*, *Asociación de Vecinos La Boca*, and others), and the environmental ombudsman of the city of Buenos Aires.
10 Meanwhile, the court referred the liability claims for personal damages to the lower courts.
11 It is worth pointing out that the mayor of the city of Buenos Aires and the governor of the province of Buenos Aires are subnational government offices with a very high level of national visibility in the context of Argentine politics. They are commonly candidates for the country's presidency or vice presidency.
12 Despite some criticism of certain aspects of the bill, opposition parties largely supported the creation of ACUMAR. The Senate approved the bill with forty-eight votes (with support from the senators of the Peronist governing coalition and almost all the opposition parties). Only two votes were cast in opposition. The House of Representatives approved the bill with 157 votes (the deputies from the government coalition and the main opposition parties, including the UCR, PRO, Federal Peronists and the Socialist party supported the bill). Only fifteen votes were cast in opposition.
13 Author's translation.
14 For more information about the Cuerpo Colegiado and its activities, see FARN's website (http://www.farn.org.ar/riachuelo/cuerpo_colegiado.html).
15 Among the members of the Cuerpo Colegiado, only the Asociación de Vecinos La Boca can be considered a local organization. This NGO is mostly comprised of local residents from La Boca, a deprived neighborhood in the city of Buenos Aires located at the mouth of the Riachuelo.

Works cited

Auditoria General de la Nación. (2006). *Cuenca Matanza Riachuelo*. Documento de Difusión. Buenos Aires.

Cámara de Diputados de la Nación Argentina. (2006). *Diario de sesiones de la honorable Cámara de Diputados de la nación Argentina*. 29a. Sesión Ordinaria; November 15, 2006.

Cámara de Senadores de la Nación Argentina. (2006). *Diario de sesiones de la honorable Cámara de Diputados de la nación*. 23. Sesión Ordinaria; October 11, 2006.

Defensor del Pueblo de La Nación. (2003). *Informe especial sobre la cuenca Matanza Riachuelo*. Buenos Aires.

Defensor del Pueblo de La Nación. (2005). *Informe especial de seguimiento sobre la cuenca Matanza Riachuelo (2003 / 2005)*. Buenos Aires.

Defensor del Pueblo de La Nación. (2012). *Relocalizaciones*. Propuesta sobre mesas de trabajo y condiciones de relocalización. Buenos Aires. Retrieved from http://www.dpn.gob.ar/areas.php?id=30&cl=24&act=view

Dixon, R. (2007). Creating dialogue about socio-economic rights: Strong-form versus weak-form judicial review revisited. *International Journal of Constitutional Law 5*, 3: 391–418.

Estrada Oyuela, R, Inhouds, E. Juliá, C., Malpartida, A, Martínez, E. and Moretton, J.(2012). *La Corte ordenó recomponer el Riachuelo. Monitoreo externo del Plan de Saneamiento de la Cuenca Matanza – Riachuelo*. Buenos Aires: Asociación de Vecinos La Boca.

FARN. Área Riachuelo. Cuerpo Colegiado. Retrieved from http://www.farn.org.ar/riachuelo/cuerpo_colegiado.html

Fundación Ciudad. (2002). *Foro Desarrollo Sostenible de la Cuenca Matanza Riachuelo*. Buenos Aires: Guía de Trabajo.

Garavito, C. R., & Franco, D. R. (2010). Un Giro en los Estudios sobre Derechos Sociales: El impacto de los fallos judiciales y el caso del desplazamiento forzado en Colombia. In

P. Arcidiácono, N. E. Yaksic, & C. R. Garavito (Eds.), *Derechos sociales: Justicia, política y economía en América Latina* (pp. 83–154). Bogota: Siglo del Hombre Editores y LAEHR.

Gauri, V., & Brinks, D. (Eds.) (2008). *Courting social justice: Judicial enforcement of social and economic rights in the development world*. New York: Cambridge University Press.

Horowitz, D. (1977). *The courts and social policy*. Washington, DC: The Brookings Institution.

Lanzetta, M., & Spósito, N. (2003). *Proceso apell Dock Sud*. Buenos Aires: PAI/CIPRA/GTZ. Not published.

McCann, M. W. (1994). *Rights at work: Pay equity reform and the politics of legal mobilization*. Chicago, IL: University of Chicago Press.

McCann, M. W. (1996). Causal versus constitutive explanations (or, On the difficulty of being so positive). *Journal of Law and Social Inquiry 21*, 2: 457–482.

Mendoza, Beatriz y otros c/ Estado Nacional y otros S/ daños y perjuicios derivados de la contaminación ambiental del Rio Matanza Riachuelo (Corte Suprema de Justicia de La Nación, June 20, 2006).

Mendoza, Beatriz y otros c/ Estado Nacional y otros S/ daños y perjuicios derivados de la contaminación ambiental del Rio Matanza Riachuelo (Corte Suprema de Justicia de La Nación, July 8, 2008).

Mendoza, Beatriz y otros c/ Estado Nacional y otros S/ daños y perjuicios derivados de la contaminación ambiental del Rio Matanza Riachuelo (Corte Suprema de Justicia de La Nación, December 19, 2012).

Nápoli, A. (2009a). Una política de estado para el Riachuelo. In M. E. D. Paola, F. Sangalli, & S. Caorsi (Eds.), *Informe ambiental anual 2009*. Buenos Aires: FARN.

Nápoli, A. (2009b). A un año de la sentencia de la corte. *Boletín de Noticias*, FARN Nro. 1. Retrieved from http://www.farn.org.ar/riachuelo/newsletter_riachuelo/NL_riachuelo1.pdf.

Nápoli, A. (2013a) El caso Riachuelo en su hora más crítica. In M. E. D. Paola, F. Sangalli, & J. Ragaglia (Eds.), *Informe ambiental anual 2013*. Buenos Aires: FARN.

Nápoli, A. (2013b). Riachuelo: A cinco años del fallo de la Corte. *Boletín de Noticias*, FARN Nro. 26. Retrieved from http://www.farn.org.ar/wp-content/uploads/2011/05/NL_riachuelo26.pdf

Nápoli, A., & García Espil, J. (2010). Recomposición ambiental de la cuenca Matanza-Riachuelo: una oportunidad histórica que aún reclama un fuerte compromiso político y más eficiencia en la gestión. In M. E. D. Paola & F. Sangalli (Eds.), *Informe ambiental anual 2010*. Buenos Aires: FARN.

Rosenberg, G. N. (1991). *The hollow hope: Can courts bring social change?* Chicago, IL: Chicago University Press.

Rosenberg, G. N. (1996). Positivism, interpretivism, and the study of law. *Journal of Law and Social Inquiry 21*, 2: 335–455.

Ryan, D. (2004). *Ciudadanía y control de gobierno en la cuenca Matanza Riachuelo*. Informe de Investigación. Teresa Lozano Institute of Latin American Studies. University of Texas at Austin. Retrieved from http://lanic.utexas.edu/project/etext/llilas/claspo/fieldreports/ryan04.pdf

Sabel, C. F., & Simon, W. H. (2004). Destabilization right: How public law litigation succeeds. *Harvard Law Review*, *117*, 4: 1015–1101.

4 Does river basin governance mean wider access to river management?

A long-term perspective from the Tennessee, Columbia, and Connecticut Rivers

Eve Vogel

Introduction

Those who care about water sustainability, democracy, and equity often believe that river governance should be organized when possible within river basins and watersheds. They argue that many people affected by water management are excluded from decision-making because they live across political borders or administrative lines and that, as a result, they bear negative consequences with little recourse. Advocates of river basin and watershed management often believe that governance systems organized within hydrological lines rather than within political and administrative borders offer the potential for a wider range of stakeholders, places, and perhaps species to influence river management decision-making—and to benefit from that management (see Vogel, 2012 for an expanded analysis). When it comes to *access*, many believe that river basin and watershed governance offer wider, more inclusive *access to decision-making* in comparison with organizing river management along local, state, provincial or national lines, and, as a result, wider and more inclusive *access to the benefits* of river management.

This chapter considers whether relationships between access and river basin organization are as positive in practice as commonly thought. To explore this question, I focus on formal river basin governance institutions—institutions organized within river basin territories rather than within more conventional political and administrative spaces, and which carry some level of autonomy and authority. In this chapter, I focus on two distinct aspects of river management: access to river management decision-making and access to river management benefits. I consider the relationship between these two kinds of access, on the one hand, and formal river basin institutions and their governance on the other hand. If proponents are right, greater access should both support and be supported by strong river basin institutions that are capable of integrating management of interrelated sectors across their basins. My questions are thus:

1 *Improved access → river basin governance?* Does wider access to river management decision-making and a push for wider access to benefits produce stronger river basin management institutions and more integrative river basin management?
2 *River basin governance → improved access?* Do stronger river basin management institutions and more integrative river basin management provide wider access to river management decision-making and benefits?

To explore these questions, I outline the histories of three multi-state river basins in the United States: the Tennessee, the Columbia, and the Connecticut Rivers (see Figure 4.1). These offer a useful comparison because there were similar efforts in the 1930s–1950s in all three basins to create river basin organizations, but they resulted in very different institutions; and because the institutions have continued since that time, with long-term effects on access to decision-making and benefits. Where not otherwise cited, material in this chapter draws from my previous work on the Columbia River (Vogel, 2007, 2012) and Connecticut River (Vogel & Lacy, 2012), as well as on a previous three-basin comparison drawing additionally on the secondary literature on the Tennessee Valley Authority (Vogel, 2015).

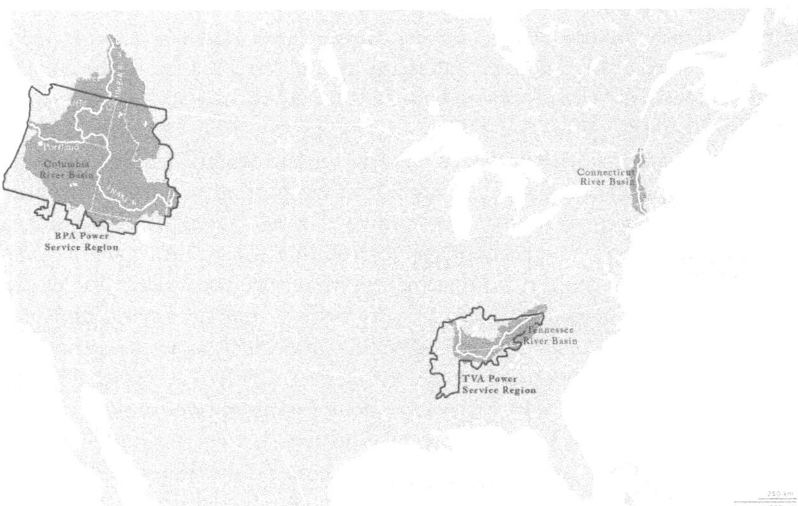

Figure 4.1 The Columbia, Tennessee, and Connecticut River Basins (basins in dark grey)

The two power service regions (black outlines) mark the territories of the two U.S. federal government agencies that were established around the Columbia and the Tennessee basins: the Bonneville Power Administration (BPA) and the Tennessee Valley Authority (TVA). The TVA was established in 1933 as a comprehensive river basin agency, while the BPA was established in 1937 as a hydropower transmission and marketing agency. No comprehensive or hydropower river basin agency was ever established around the Connecticut River basin.

Source: Cartography by Jessica McDaniel, Gregory Woolston, and Eve Vogel, using data from Natural Resources Canada, Northwest Habitat Institute, USGS EROS, Esri Data & Maps, ICBEMP (c/o USFS Pacific NW Research Stn and USF&W), BPA, TVA, and Natural Earth.

Background: thinking critically about river basin institutions, management, and access

Despite proponents' enthusiasm, a growing body of work shows that organizing water management within river basins does not guarantee universal access to decision-making, nor does it provide equal access to river-related resources. Switching to a watershed-based governance system shifts priorities, participants, and practices, but, like any other kind of governance, it furnishes some participants with better access to decision-making than others.

In previous papers focusing on river basin management of the Columbia, Connecticut, and Tennessee River Basins (see especially Vogel, 2012, 2015), I found no clear linear relationship between stronger river basin management and its frequently claimed beneficial effects: river basin management in these rivers did not necessarily produce more holistic or inclusive decision-making, more balanced management, more even distribution of benefits and costs, or improved sustainability of biophysical systems and ecological productivity. Part of the failure to meet expectations, I argued, is that the expectations of positive effects derive from the belief that river basin management institutions can be free from the political conflicts and the constraints of conventional political jurisdictions. I showed instead that institutions working within a river basin territory interact with, and often depend on, institutions operating within more traditional government, political, and administrative spaces.

This chapter's focus on the issue of access illuminates one reason the interaction between river basin institutions and traditional governmental and administrative spaces matters so much. Governments and administrations operating within traditional, non-basin political spaces play a vital role in providing access to decision-making in river basin management. This is so for two central reasons: these governments and administrations provide direction, authority, funds, and other essentials in the establishment of river basin management institutions; and they provide alternative routes to influence river basin decision-making, via their ongoing authorities and funding mechanisms, when river basin institutions themselves become closed.

1930s–1950s: improved access → river basin governance?

This section considers how political access shapes the creation of river basin institutions. Does wider access to river management decision-making and a push for wider access to benefits produce stronger river basin management institutions and more integrative river basin management?

The Tennessee River: an unusual opportunity to build strong integrative river basin management

Created in 1933, the Tennessee Valley Authority (TVA) was the first in a series of valley authorities around the world. No other valley authority was ever created in the United States. What kinds of political interests and access to decision-making shaped the TVA?

Advocates for creating the TVA included a range of progressive activists and thinkers who aimed to develop the river with dams, spread the river benefits of jobs, electricity, and navigation throughout the basin, conserve forests and soils in the hillslopes, create model communities; and educate farmers (Lilienthal, 1953). Their access to influence policy came mainly through the U.S. federal government in a moment of unmatched political opportunity.

In the early 1930s, the world was in the depths of the Great Depression. The new president, Franklin D. Roosevelt, or FDR, promised a "New Deal." The federal government would help to rein in the abuses of speculative capitalism, and, using its own moneys as well as its planning and administrative capacities, would restart more reliable business, industry, and agriculture, build infrastructure, and put people to work. Roosevelt pushed for a river valley authority as an integrative way to achieve these goals across an entire region (Leuchtenburg, 1963).

Why the Tennessee River? A federal dam and two nitrate plants had been constructed here during World War I and could function as seeds of a regional river development program. The Tennessee offered more hydropower potential than any other river in the American East. Because the region was poor, it represented an opportunity to advance progressive goals. Elites and boosters from the region also supported the creation of the TVA because it would bring federally funded development. Finally, the president and Congress were eager to send federal dollars to the Tennessee Valley because it was in the American South, which had been essential to FDR's Democratic electoral landslide (Hargrove, 1994; McCraw, 1971).

However, even historic political access and geographic opportunity were insufficient to win a full-fledged river basin agency. Equally crucial were the unusual challenges faced by the powerful opponents of river basin agencies—electric companies and investors, and existing federal agencies. In the 1920s and 1930s, electricity was a lucrative, cutting-edge industry. Owners, managers, and investors in the electrical industry faced the prospect of a sudden gush of cheap federal hydroelectric power, especially since the TVA, under federal doctrine, would be directed to sell its power preferentially to publicly and cooperatively owned utilities instead of to investor-owned ones. In the late 1920s, though, two federal reports exposed the electrical industry's propaganda, and the stock market crash had been linked to its speculative investment schemes. FDR himself had railed against the "power trust" during the 1932 election. In 1933, these usually powerful opponents found themselves with little political access to stop unwanted policy (McCraw, 1971).

Less publicly visible but perhaps even more powerful opponents were operating within the federal government. At least two federal departments enjoyed a longstanding presence along the Tennessee River and its hillslopes: the Department of War, because its Army Corps of Engineers (Corps), was the main builder of dams, navigation channels, and levies in the region, and the Department of Agriculture, because of its work with farm programs and forests. Both stood to lose responsibilities to an upstart, multi-sector regional agency. But, like the electrical industry, these departments had little ability to stop the leading initiatives of an overwhelmingly popular president during his first few months in office (Leuchtenburg, 1952, 1963).

Thanks to these rare circumstances of political access—unusually strong access for progressives and regional elites, and unusually weak access for the private electrical industry and existing federal agencies—the TVA was created in summer 1933, vested with considerable authority and autonomy, and empowered to integrate management of multiple sectors across its basin.

The Columbia River: compromising to form a river basin power agency

After the creation of the TVA, many proposals for additional river basin authorities followed. Like the Tennessee, the Columbia Basin offered particular opportunities for federal intervention. In 1933, FDR approved two massive federal dams, Bonneville and Grand Coulee, on the Columbia and, by the mid-1930s, they were under construction. The Columbia offered greater hydropower potential than any other river in the country, and, because the Pacific Northwest was relatively undeveloped and features an arid interior, the promise of irrigation was a major benefit (BPA, 1981; Tollefson, 1987).

Progressives and regional boosters in the Pacific Northwest were enthusiastic about the potential of a river basin agency in the Columbia to accelerate regional development. The Pacific Northwest was also home to a strong movement in favor of publicly owned electrical power, a movement led by farm, labor, and some industrial interests. In 1930, advocates in the states of Washington and Oregon had used citizen initiatives to pass landmark laws enabling residents to form county-based public utility districts. Now, locals clamored for a new river basin agency that would produce and distribute to these public utilities inexpensive, widely available federal hydropower (BPA, 1981; Tollefson, 1987).

After the TVA's creation in 1933, though, opponents of river basin agencies marshaled their forces, re-establishing their access to influence policy. Electric companies and investors launched federal lawsuits, ran national media campaigns, lobbied Congress, and backed political candidates who promised to advance their interests. The Secretaries of War and Agriculture fought proposals for all new valley authorities. Opponents in the Northwest mobilized too. Leaders of Portland, Oregon, lambasted the idea that electric power from the Bonneville Dam, which was under construction nearby, might be shared with the entire region rather than sold locally. Politicians in the upper-river states of Idaho and Montana opposed a regional agency that might make claims on their water, lower the price of electricity so that it could no longer subsidize irrigation, or decrease the chances of receiving their own new federally-funded dams (Leuchtenburg, 1952; McCraw, 1971; Vogel, 2007).

In response, FDR delayed. He waited until after 1936, when the Supreme Court ruled in the TVA's favor and he won another election by a landslide. By this time, opponents were strong enough to kill proposals for a series of new valley authorities (Leuchtenburg, 1952; McCraw, 1971).

Nevertheless, a Columbia Basin agency had one crucial impulse propelling it forward. Bonneville Dam was due to come on line near the end of 1937, and no legislation had passed yet authorizing its hydropower to be delivered and sold. Politicians from Oregon and Washington called for Bonneville's power to be made available as quickly as possible. This required Congressional action.

The president insisted on a new federal agency. However, he offered a significant compromise. The Corps would continue to operate Bonneville Dam and its navigation locks, and the U.S. Bureau of Reclamation would continue to operate Grand Coulee Dam and deliver its irrigation waters. The new agency would be responsible only for transmitting and selling electric power. At first, it was given responsibility only for Bonneville Dam, but later, as other federal dams were authorized and built on the Columbia River and its tributaries, the agency was given responsibility for transmitting and marketing the power from these dams as well (BPA, 1981; Vogel, 2007).

Thus was the Bonneville Power Administration born and shaped. Crafted from the same vision as the TVA, it carried significantly less authority, with much greater dependence on other agencies—but still with an ambitious mission to spread the benefits from federal Columbia River dams region wide (BPA, 1981; Tollefson, 1987; Vogel, 2007).

Connecticut River: resisting river basin governance in favor of fragmentation

Though small in comparison to the Columbia and even the Tennessee, the Connecticut River is the largest in New England. New England, in turn, is a small region with a strong identity and a large place in American history, economy, and politics.

Advocates and opponents of river basin agencies lined up differently in New England than in other regions, and they had distinct kinds and levels of access to decision-making. Electrical industry executives remained dominant players in New England's economic and political affairs. Regional elites thus vehemently opposed federally produced electric power because it would undercut electricity prices and profits for utilities and investors. They also harbored particular antipathy toward the New Deal. Unlike the South and the Pacific Northwest, New England was an already industrialized region, indeed, a region that was starting to de-industrialize. From their point of view, the New Deal diverted taxes from long-successful New Englanders to fund development in competitor regions. Even if the New Deal was offering them a fraction of the same investment, they set themselves firmly in opposition (Leuchtenburg, 1953).

New England progressives, especially in the upper-river state of Vermont, also took a different stance on federal dams from their counterparts along the Tennessee and Columbia Rivers. They defended interests almost un-acknowledged in public debates elsewhere in the 1930s, namely the interests of residents and lands of river valleys that would be flooded by reservoirs. New England's river valleys had long been relatively densely settled, and amenity tourism was already an important economic sector (Judd, 1979; Leuchtenburg, 1953).

Still, New Englanders had limited access to political decision-making in Washington, DC. Not only did they oppose the New Deal, their political representatives were mostly Republicans, while Roosevelt and a large majority in Congress were Democrats. New Englanders worked indirectly. They gained seats on federally appointed regional planning boards and water development commissions, and led these groups to recommend against federal river basin development. They also lobbied their members of Congress to oppose federal river basin development and replaced them when they did not (Leuchtenburg, 1953).

Even so, New Englanders could not deflect all arguments for federal river development. Major river floods hit the region in 1927, 1936, 1938, on New Year's in 1948–1949, and in 1955. Each flood prompted popular demand for flood control. There were also growing calls to reduce chronic water pollution from the many factories along the region's rivers. Large-scale federal river development could help both causes (Leuchtenburg, 1953; Parkman, 1978).

The region's elite, who hailed mainly from downstream states, fought federal river development proposals back to a smaller set of single-purpose, flood control-only dams. Then, to limit the role of federal authority in the region, they began to develop alternatives: interstate compacts and a voluntary watershed association to coordinate flood control and lead on pollution abatement. For these alternatives, though, they needed the support of New England's upper-river states.

The anti-dam populists defending up-river valleys leveraged impressive political access and began to threaten even a modest program of federal flood control dams. In 1937, a leading progressive Republican, George Aiken, became governor of Vermont; in 1941, he became a U.S. Senator. Under his leadership, Vermont challenged the federal government's power to take land using federal eminent domain. Aiken led newspapers and politicians around the country to cry out against federal intrusion and usurpation of state and local autonomy. When the Corps began to survey lands for possible reservoirs, Vermonters used their local governments and state agencies, encouraged by newspapers from around the state, to cite the surveyors for trespassing, deny them permits to buy explosives, and threaten to take away their preferred gas rations (Leuchtenburg, 1953).

Legislators and businessmen from lower-river states strategized and negotiated. They used the threat of broader federal intervention to cajole their upper-river counterparts to support interstate or citizen agreements. In return, they agreed to fewer flood control dams and offered seats at regional decision-making tables. Together, New Englanders shepherded through a New England Interstate Water Pollution Control Commission, approved by Congress in 1949, and a Connecticut River Valley Flood Control Commission, approved by Congress once a Republican finally won the presidency in 1953. In 1952, a group of citizens and business leaders also formed the nonprofit Connecticut River Watershed Council, which ran media campaigns and citizen clean-ups (Leuchtenburg, 1953; Vogel & Lacy, 2012).

Twenty years after the TVA's establishment, the Connecticut River finally had three very limited basin-wide institutions. None of these would have the power and authority to manage the basin in an integrated fashion, as the TVA could do, or as the BPA could do in concert with the Corps and the Bureau of Reclamation. Investor-owned electric companies continued to own and run the vast majority of hydropower dams on the mainstem and tributaries, and they did it without much coordination, and little attention to river basin management or regional development. River basin governance remained fragmented in the Connecticut basin.

1950s–present: river basin governance → improved access?

By the 1950s, a strong, relatively autonomous river basin organization had formed in the Tennessee River basin; a river basin hydropower agency had been formed along the Columbia River, dependent on other agencies and political support; and there was no strong or integrated river basin organization in the Connecticut River. How did the varying institutions of river basin management in these three basins, and their different levels of integration, subsequently affect access to river management decision-making and river benefits? To analyze these questions, I outline four parallel time periods and themes in each of the basins:

- 1940s–1960s: institutional consolidation and infrastructural expansion
- 1970s: nuclear expansion, spiraling costs, and environmental challenges
- 1990s: regional electric power versus competitive energy markets
- 2000s–present: recent events and long-term legacies

The Tennessee River: wide benefits, sharp impacts, limited access to decision-making

1940s–1960s: institutional consolidation and infrastructural expansion

Thanks to its strong authorities and integrative approach, the TVA was able to spread access to river benefits widely. In the beginning, the TVA used its authority to acquire land, build dozens of dams, and market electric power. It wrested control over electric power provision from the private sector. Electric service was widened dramatically while electric rates plummeted. The TVA's dams extended the reach of navigation upriver to Knoxville, Tennessee. In the agency's early years, the TVA also made fertilizer widely available and offered technical expertise in soil and forest conservation to many farmers and land managers.

However, there were high social and environmental costs, and the least powerful had little voice. The TVA's dams flooded massive areas of fertile valley bottomland, displacing over 50,000 people and greatly altering river ecosystems. Faced with opposition from southern politicians, the TVA backed off from pressing for racial equity and equal representation for tenant farmers. By World War II, the TVA had honed its focus to economic development and electric power provision. In the postwar period, with few dam sites left, the TVA built numerous coal-fired power plants. Rural areas faced rapid out-migration, and benefits concentrated in the basin's cities.

In the 1950s, the TVA region's congressional representatives helped the agency win even greater decision-making independence. Most importantly, in 1959, the TVA won the right to finance construction of new power plants itself.

1970s: nuclear expansion, spiraling costs, and environmental challenges

With greater autonomy, the TVA was able to restrict access to decision-making for less powerful challengers. Starting in the 1960s and continuing into the 1970s, the TVA plowed through several major controversies, despite massive impacts and significant protests. First, it strip-mined coal on an accelerated scale. Next, it took lands from people along the last free-running, biologically rich stretch of the Little Tennessee River in order to create a recreation reservoir and sell the lands to developers as an economic development opportunity. Even the Endangered Species Act and the Supreme Court could not stop the boondoggle. Finally, despite rising costs and concerns, it pushed forward its massive program of nuclear power, the largest in the country (Hargrove, 1994; McCraw, 1995; Murchison, 2007; Plater, 2013).

1990s: regional electric power versus competitive energy markets

In the early 1980s, TVA's mounting costs came together with an economic recession. Escalating electricity prices made even the region's business leaders and congressional representatives—people with access to TVA decision-making—critical. At the same time, there was a growing national and international drive for electric deregulation, privatization, and competition. A major corporate executive brought in to chair the TVA in 1987 instituted widespread layoffs, and, by the 1990s, helped transition the TVA "from a regional bureaucracy to a more competitive, more efficient, and more focused corporation" (Ezzell, 2008, p.20; FN 61). But the agency was not privatized, nor were its markets deregulated. These measures would have stopped the agency from conducting its most profitable business, which was also its most important distribution of benefits: selling cheap electric power across its region. Despite being generally pro-free market, conservative southern Republicans, the region's congressional representatives wanted the region to continue enjoying access to TVA's power, so they defended the importance of federal ownership of the TVA.

2000s–present: recent events and long-term legacies

With limited access to TVA decision-making, challengers have had to go to outside jurisdictions in order to force change on the agency. In recent years, they have filed environmental lawsuits and won new clean air regulations. The TVA has also faced flat or declining electric demand, low natural gas prices, and looming changes in electric utility markets. Eighty years after its creation, the agency is finally widening access to its decision-making and beginning to address a broader range of benefits.

In 2011, the TVA created a new Integrated Resource Plan and committed to "lead the nation in improved air quality and increased nuclear production and to lead the Southeast in increased energy efficiency." In 2015, it revised the plan through perhaps its most participatory planning process to date. It has finally backed off from further nuclear and coal ambitions. Still, its primary missions remain the same as in the 1940s: cheap power and regional economic development. Even in 2015, the agency declines to give significant support to renewables or conservation, citing potential electricity rate increases and perhaps fearing the wrath of its congressional representatives. Problems in the river and riverside environments receive secondary attention (Casey, 2015; Tennessee Valley Authority, 2011, 2015).

The Columbia River: wide access to cheap electric power, environmental mitigation, and participatory decision-making

1940s–1960s: institutional consolidation and infrastructural expansion

In contrast to the TVA, the BPA had to coordinate with, and often placate, multiple federal, state, private, and local entities. Top among these were the federal Army Corps of Engineers and U.S. Bureau of Reclamation, which owned and operated the dams that generated the BPA's electric power. Like the TVA, the BPA also worked closely with its power service region's congressional delegation. And the BPA depended on its customers: the region's hundred-plus local public and cooperative utilities, which constituted the BPA's "preference" customers; a number of large industries that were vital bulk power customers; and finally, investor-owned utilities, which remained stronger in the Pacific Northwest than in the Tennessee region and which tolerated the BPA if it sold them cheap electric power. These institutions retained strong access both to influence BPA policy and to claim Columbia River benefits.

In its early years, the BPA helped a range of local and state areas to access benefits—new dams, industrial investment, and other resources. In the 1960s, the Columbia River Treaty extended river management northward into Canada, and the Pacific Intertie extended power distribution southward into California. To win these agreements, BPA and Army Corps officials agreed to coordinate decision-making, financing, and river management, and to share benefits with a host of old and new partners. Working with its partners, the BPA coordinated river flows and acted as a central broker and coordinator in the development of an increasingly participatory river management system linked to an ever-wider distribution of benefits.

1970s: nuclear expansion, spiraling costs, and environmental challenges

Running out of dam sites prompted the BPA, like the TVA before it, to turn to coal and nuclear power. However, the BPA ran out of dam sites twenty years after the TVA did. Only in the late 1960s did the BPA and its partners launch a coal and nuclear energy expansion program. This timing, together with the BPA's limited authorities, meant that challengers in the contentious 1970s were able to wield much more influence over river management on the Columbia River than did their counterparts on the Tennessee—and to claim new kinds of benefits as well.

Unlike the TVA, the BPA had never received authority to build its own power generation plants. Thus, during the coal and nuclear expansion period of the late 1960s and 1970s, it served as coordinator and financial backer when dozens of local publicly and cooperatively owned utilities joined together to collaborate in issuing investment bonds in five nuclear power plants. In 1970 and 1975, the cities of Eugene, Oregon, and Seattle, Washington—owners of the largest public utilities in their respective states—voted to suspend their involvement in nuclear power and called instead for energy conservation. Their pull-out threatened the BPA-backed nuclear program and presented the BPA with a strong push toward energy conservation, a push made stronger by the 1973 global energy crisis. By the late 1970s, BPA's rickety financing system was ruled illegal, and cost overruns—similar to those the TVA had weathered—were sending the Northwest's public nuclear power program toward default.

Desperate, Pacific Northwest utilities advanced a bill in Congress to augment the BPA's authority. With a bill in the federal Congress, challengers in the region had a unique opportunity to access river and agency policy decision-making. Conservation advocates inserted what would become the world's first mandate to prioritize energy conservation. Native American tribes and other fish advocates added a Columbia River Fish and Wildlife Program to the bill. The 1980 Northwest Power and Conservation Act institutionalized these measures and widened public access to decision-making regarding the Columbia River and the Pacific Northwest power system.

1990s: regional electric power versus competitive energy markets

In the 1990s, three new pressures came down on the BPA: a push for energy deregulation, competing low-cost power from independent natural gas generators (themselves a product of early deregulation), and new legal mandates for protecting native salmon. The BPA had already begun to address the issue of energy conservation, a goal that justified deregulation elsewhere and had already brought in wide public participation. Now, a range of stakeholders helped to stabilize the agency with new kinds of contracts and sales agreements. Gas prices then went up again, the West Coast power crisis hit in 2001, and Northwestern policy makers backed off from further deregulation. When a number of Columbia River salmon runs were listed under the Endangered Species Act, the BPA responded with a world-class environmental mitigation program, doling out hundreds of millions of dollars each year for fish and wildlife programs.

2000s–present: recent events and long-term legacies

Today, a range of groups and interests continue to have access to influence Columbia River decision-making and receive benefits. While power generation remains the primary goal of river management, proceeds from that generation now fund a large set of benefits and beneficiaries: energy conservation, low-income weatherization, cutting-edge fish research, fish hatcheries and fish passage, and land acquisition for habitat conservation. Among the recent innovations: since 2008, a large fund has been dedicated to Columbia River Native American tribes for Columbia River fish and wildlife. Native leaders have viewed this effort as respecting their sovereignty in a way many previous efforts did not (Anon, 2008).

The Connecticut River: uneven access to benefits and fragmented decision-making

1940s–1960s: institutional consolidation and infrastructural expansion

In the Connecticut River's postwar period, the three limited basin-wide institutions that had emerged from the push for a river basin authority—the interstate pollution and flood control commissions, and the non-governmental watershed council—kept their focuses narrow. The interstate pollution commission and the Connecticut River Watershed Council mounted campaigns to reduce pollution. The Army Corps of Engineers built thirteen flood control dams, each constructed as low as possible, while sufficiently protecting mainstem river locations from major flood damage. The interstate flood commission worked out costs and payments between the states. Meanwhile, various investor-owned utilities managed their own dams on the mainstem and large tributaries. Without a strong river basin agency, Connecticut River management remained fragmented, its ambitions and achievements limited.

1970s: nuclear expansion, spiraling costs, and environmental challenges

Between 1965 and 1980, a somewhat more integrative institution emerged in the Connecticut River basin, thanks to a renewed but modest federal government initiative to support river basin planning: the New England River Basins Commission. Planners advised a range of projects and issues but lacked independent authority to act.

Yet the River Basin Commission and other river stakeholders lacked access to influence the operations of hydropower dams owned and run by the investor-owned electric power sector. In the 1970s, New England electric utilities mounted a large nuclear buildup campaign, as had utilities in the Tennessee and Pacific Northwest regions, and they also built a giant pump-storage station on the Connecticut River to help meet peak demand (MacAvoy & Rosenthal, 2005). The New England River Basins Commission's main involvement was to accept the pump-storage station and to facilitate discussions about the downstream warming effects of the Vermont Yankee nuclear plant (Foster, 1984).

Before the River Basins Commission was de-funded in 1980 by the Reagan Administration, it helped birth the Connecticut River Atlantic Salmon

Commission, established in 1983. The salmon commission advocated for fish passage through the river's lower dams, established fish hatcheries in the middle and upper rivers, and supported ongoing research (Foster, 1991). Still, its access to avenues to influence policy was limited. To improve fish passage and flow conditions, advocates had to wait until the licenses of individual dams ran out—every thirty to fifty years, with the dates depending on the dam and its particular cycle—and then opportunistically try to influence new license requirements.

1990s: river basin organization and regional electric power versus competitive energy markets

In the 1990s, New England progressives, having no need to protect a large regional public electric power agency like the TVA or BPA, joined the push for energy deregulation. They saw competitive markets as a route to more efficient energy use, and, in a region with some of the highest electric rates in the nation, they hoped competitive markets might make electricity more affordable. Thanks in part to their political access and savvy, Massachusetts deregulated its electric utilities in the 1990s, and a New England-wide "independent system operator" was launched in the 2000s to manage newly competitive wholesale electric power markets. From the outset, energy conservation was a goal in these competitive markets (Massachusetts Division of Energy Resources, 1996; Polestar Communications & Strategic Analysis, 2006).

Following implementation of these regulatory reforms, many of the Connecticut River hydropower dams were sold by utilities to non-utility electric generation companies. In the wake of this change, different hydropower generators were no longer at liberty to communicate with one another regarding plans for flow releases, as this was deemed anti-competitive collusion. Basin-wide management arguably became even more challenging. Also, in some circumstances, Connecticut River flows became newly subject to price changes in the electric spot market. Beyond advocating for restrictions within the licenses of individual dams, it was not possible for those interested in river management to exert influence over these electric market effects on the river.

2000s–present: recent events and long-term legacies

Attempts to advance basin-wide Connecticut River conservation persist today, but advocates have limited access to funding and authority and little ability to influence large-scale management coordination. In 1997, Congress created a national wildlife refuge consisting of the entire Connecticut River basin, but the refuge has limited funding and owns little land. Salmon restoration was finally abandoned after 2011, when Tropical Storm Irene destroyed the major hatchery. The Connecticut River Atlantic Salmon Commission has renewed focus on other species but also operates with limited funding and power. Unlike the Tennessee and Columbia basins, electric power sales in the Connecticut basin fund conservation and mitigation programs only indirectly and piecemeal. In the 1990s, 2000s, and 2010s, several of the major Connecticut River hydropower projects came up for relicensing, and stakeholders were able to win more fish passage as well as a mitigation fund of several million dollars for various habitat improvement projects (see e.g. Tripp, 2005). However, these initiatives are not coordinated, nor are they basin wide.

The failure of basin management on the Connecticut River has produced mixed consequences ecologically, economically, and in terms of access to river benefits. On the one hand, much less land was permanently inundated, sparing numerous Vermont communities and valleys, while on the other hand, the river produces less hydropower than it might. While mainstem flood control is effective, many tributaries can still swell to destructive proportions during high-flow events. On an annual and seasonal basis, dams do not overly alter river flows, and federal flood control dams often provide natural tributary flows to aid in fish restoration. In short, failure to coordinate seems to have staved off large-scale interventions that could have supplied benefits to many people while also devastating the river ecosystem and its human communities' ability to fully access and enjoy the benefits of natural rivers and valleys.

How do access and river basin governance influence one another?

This section reviews the two questions with which I began.

Improved access → river basin governance?

Does wider access to river management decision-making and a push for wider access to benefits produce stronger river basin management institutions and more integrative river basin management? Based on the histories of these three rivers, the answer is a qualified, partial, or sometimes yes.

As both advocates and detractors believe, increased political access—in particular for those who oppose more conventionally organized agencies and powers, and who push for access to benefits for wider areas and populations—may be crucial in establishing formal river basin organizations. This is a central lesson of the TVA and BPA stories. Equally important in these stories is the importance of diminished political access for existing agencies and powers during the initial phase of river basin organizations.

However, widened access to decision-making does not necessarily lead to support for river basin governance institutions. It depends on the kinds of river benefits desired by access-empowered constituents and advocates. In the three basins featured, advocates of underserved people and interests supported river basin governance when they wanted benefits from river development—hydropower, navigation, flood control, and irrigation. If they valued fisheries, floodplains, valleys, or other resources of un-dammed rivers—as they did in the Connecticut Basin—they opposed river basin governance. In the Connecticut basin, widely harnessed democratic access helped stop a strong river basin organization.

These three histories also suggest that the preferences of regional elites and regional elites' access to national government policy decision-making can determine the development or obstruction of a river basin organization. Elites in relatively undeveloped regions may be interested in a river basin organization if it will bring large outside investments. However, elites in relatively developed regions may oppose river basin organizations in order to protect their own existing earnings against a developmentalist state that will divert funds to national peripheries.

River basin governance → improved access?

Do stronger river basin management institutions and more integrative river basin management provide wider access to river management decision-making and benefits? Based on the histories of these three rivers, the answer is generally yes with regard to accessing benefits and generally no with regard to accessing decision-making. Both answers come with a qualification, however.

River basin governance → access to benefits?

Generally, yes. Strong and integrated river institutions can help spread access to river benefits. The TVA and BPA successfully increased electric power production, lowered rates, and spread service across their power service regions. Not so in the Connecticut Basin, which lacked a river basin agency.

But this observation must be heavily qualified. The TVA and BPA widened access to benefits, but mainly to a narrow set of benefits—in both cases especially hydropower—and this narrow set of benefits was often maximized at high cost to other river resources such as valley lands, fisheries, and the people who depended on these resources. Indeed, costs were especially high and rarely mitigated in the TVA, where river basin management began as the most integrative and remains the strongest in terms of authority and autonomy. In the Columbia River, the BPA's greater dependence on other agencies and political support led to much earlier accommodation of different kinds of river resources and users. Even here, though, the BPA tended to accommodate non-power interests through mitigation moneys more than through fundamentally changing river management.

River basin governance → access to decision-making?

Generally, no. In these three basins, stronger river basin institutions did not provide greater access to river-related decision-making. Indeed, a strong integrated river basin organization—the TVA—actively impeded access to river management decision-making.

In the cases of the Columbia and Connecticut Rivers, a range of activists and challengers have enjoyed far more access to river basin decision-making than those in the Tennessee River. In the case of the Connecticut River, because institutions and management remain fragmented, influencing decision-making tends to mean influencing only some small part of the river's management. In the absence of a strong river basin agency, however, fragmented power may win the day. Among the three basins, only in the Connecticut River Basin were activists able to stave off large-scale development and its attendant impacts.

The story of the Columbia is unique in a different way. Its challengers, representing a range of interests, have succeeded in influencing large-scale policy in key ways. The BPA represents a middle ground of a fairly strong but not autonomous river basin institution, which is deeply dependent on other agencies, groups, political representatives, and policies. The characteristics of this middle-ground institution seem to maximize access to large-scale river basin decision-making.

Conclusion

The histories of these three basins suggest that access and formal river basin governance institutions influence one another but in more complex and ambiguous ways than advocates often predict. A key to understanding this complexity is recognizing that river basin organizations must be constructed through other jurisdictions, other political spaces, even though, when this happens, the organizations often aim to overcome the constraints and political pressures of those more conventional jurisdictions. If river basin organizations succeed in gaining considerable autonomy, they can become politically inaccessible to outside interests. The TVA's history illustrates this problem. For challengers to access decision-making, and in particular to succeed in widening the kinds of benefits delivered from a river, they may have to work through other policy arenas, arenas generally organized in more traditional governmental spaces. Thus, the continued role and power of conventional jurisdictions—either in an institutionally and geographically fragmented system like the Connecticut or an institutionally interdependent system like the Columbia—remain important to ensure continued access within river basin governance.

Works cited

Anon. (2008, May 9). Tribes, federal agencies sign "Columbia Basin Fish Accords" resulting from BiOp remand. *Columbia Basin Bulletin*. Retrieved from http://www.cbbulletin.com/Free/273910.aspx

BPA (Bonneville Power Administration) (1981). *Columbia River power for the people: A history of policies of the Bonneville Power Administration*. Portland, OR: Bonneville Power Administration.

Casey, T. (2015, July 14). TVA puts integrated resource plan where the sun don't shine. *Johnson City Press*. Johnson City, TN. Retrieved from http://www.johnsoncitypress.com/News/2015/07/13/TVA-puts-Integrated-Resource-Plan-where-the-sun-don-t-shine-1.html

Ezzell, P. B. (2008). *TVA photography, 1963–2008: Challenges and changes in the Tennessee Valley*. Jackson, MS: University Press of Mississippi.

Foster, C. H. W. (1984). *Experiments in bioregionalism: The New England river basins story*. Hanover, NH: University Press of New England.

Foster, C. H. W. (1991). *Yankee salmon: The Atlantic salmon of the Connecticut River*. Cambridge, MA: CIS.

Hargrove, E. C. (1994). *Prisoners of myth: The leadership of the Tennessee Valley Authority, 1933–1990*. Princeton, NJ: Princeton University Press.

Judd, R. M. (1979). *The New Deal in Vermont: Its impact and aftermath*. New York: Garland Pub.

Leuchtenburg, W. E. (1952). Roosevelt, Norris and the "Seven Little TVAs." *The Journal of Politics, 14*, 3, 418–441.

Leuchtenburg, W. E. (1953). *Flood control politics: The Connecticut River Valley problem, 1927–1950*. Cambridge, MA: Harvard University Press.

Leuchtenburg, W. E. (1963). *Franklin D. Roosevelt and the New Deal, 1932–1940*. New York: Harper & Row.

Lilienthal, D. E. (1953). *TVA: Democracy on the march*. New York: Harper.

MacAvoy, P. W., & Rosenthal, J. W. (2005). *Corporate profit and nuclear safety: Strategy at Northeast Utilities in the 1990s*. Princeton, NJ: Princeton University Press.

Massachusetts Division of Energy Resources. (1996). Power choice: A plan to deregulate elements of the electric industry in Massachusetts. Boston, MA. Retrieved from https://ia700708.us.archive.org/33/items/powerchoiceplant00mass/powerchoiceplant00mass.pdf

McCraw, T. K. (1971). *TVA and the power fight, 1933–1939*. Philadelphia, PA: Lippincott.

McCraw, T. K. (1995). The hubris of the engineers. Review of Edwin C. Hargrove, *Prisoners of myth: The leadership of the Tennessee Valley Authority, 1933–1990*, Princeton University Press, 1994. *Technology and Culture, 36*, 4, 1007. http://doi.org/10.2307/3106923

Murchison, K. M. (2007). *The Snail Darter case: TVA versus the Endangered Species Act*. Lawrence, KS: University Press of Kansas.

Parkman, A. (1978). *Army engineers in New England: The military and civil work of the Corps of Engineers in New England, 1775–1975*. Waltham, MA: U.S. Army Corps of Engineers, New England Division.

Plater, Z. J. B. (2013). *The Snail Darter and the dam: How pork-barrel politics endangered a little fish and killed a river*. New Haven, CT: Yale University Press.

Polestar Communications & Strategic Analysis. (2006, September). A review of electricity industry restructuring in New England. Boston, MA: The New England Energy Alliance. Retrieved from http://www.ksg.harvard.edu/hepg/Papers/NEEA_0906.pdf

Tennessee Valley Authority. (2011). Integrated resource plan: TVA's environmental & energy future. Retrieved from http://nnsa.energy.gov/sites/default/files/nnsa/08-14-multiplefiles/TVA%202011w.pdf

Tennessee Valley Authority. (2015). Integrated resource plan, 2015 Final Report. Knoxville, TN: Tennessee Valley Authority. Retrieved from https://www.tva.com/file_source/TVA/Site%20Content/Environment/Environmental%20Stewardship/IRP/Documents/2015_irp.pdf

Tollefson, G. (1987). *BPA and the struggle for power at cost*. Portland, OR: Bonneville Power Administration.

Tripp, N. (2005). *Confluence: A river, the environment, politics, & the fate of all humanity*. Hanover, NH: Steerforth Press.

Vogel, E. (2007). The Columbia River's region: Politics, place and environment in the Pacific Northwest, 1933–present (Unpublished doctoral dissertation). Eugene, OR: University of Oregon.

Vogel, E. (2012). Parcelling out the watershed: The recurring consequences of organising Columbia River management within a basin-based territory. *Water Alternatives*, 5, 1, 161–190.

Vogel, E. (2015). River basins versus politics?: Interactions, patterns and consequences. In E. S. Norman, C. Cook, & A. Cohen (Eds.), *Negotiating water governance: Why the politics of scale matter* (pp. 59–78). Farnham, Surrey, UK: Ashgate.

Vogel, E., & Lacy, A. (2012). New Deal versus Yankee independence: The failure of comprehensive development on the Connecticut River, and its long-term consequences. *The Northeastern Geographer*, 4, 2, 66–94.

5 Shifting currents in water diplomacy

Negotiating conflict in the Danube and Nile River basins

Catherine M. Ashcraft

Introduction

Rivers, according to the ancient Romans, are *limes*, frontier spaces for interaction and dialogue that divide and join different groups and are subject to inherent conflict. Rivers that cross international boundaries connect countries' ecosystems, people, and economies to one another. At times this interaction fosters competition and at other times collaboration. Water users, managers, and diplomats need practical strategies to productively manage both, especially as climate change makes already complex social–ecological systems even less predictable and more vulnerable to chronic disasters, such as droughts, floods, and ecosystem changes.

Environmental conflict emerges when two or more interdependent parties perceive differences in their interests, positions, or goals concerning how to share or preserve a limited natural resource or its benefits, or how to create something new together, and yet believe they cannot achieve their goals simultaneously (Fisher, 2014; Pruitt, Rubin, & Kim, 2003). A common myth is "all conflict is bad" and should be reduced or avoided. However, conflict, like water, is not inherently good or bad and has the potential to be both destructive and productive. Unmanaged conflict can escalate, dampen communication, and contribute to negative emotions, but productively managed conflict is also a source of creativity and change, which opens opportunities to challenge the status quo (Lewicki, Saunders, & Barry, 2015). In practice, strategies to reduce environmental conflict's destructive impacts and take advantage of its productive potential are not mutually exclusive and are often used together. This chapter's analysis of the opportunities and risks posed by three common conflict management strategies, as well as their combined effects, provides practical insights for water users, managers, and diplomats engaged in critical water negotiations, and conflict resolution theorists seeking to improve the quality of deliberation.

When conflict is perceived negatively and negotiators want to reduce its destructive impacts, they often search for "low-hanging fruit," low-risk agreements they can reach with relatively little effort. One strategy is to narrow the range of issues for negotiations by building only on noncompetitive similarities, shared interests over which negotiators do not compete despite their differences in other areas, and avoiding issues about which they disagree. For example, even countries that disagree over water rights share an interest in reducing the number of people within their countries without access to sanitation. A second strategy is to narrow the range of participants through "party arithmetic," by limiting the ability of parties with dissenting interests to participate in making water policy decisions. For example, countries making decisions to protect and enhance biodiversity could exclude input from other interest groups impacted by these decisions, such as hydropower generators (Lax & Sebenius, 1991; Mnookin, Peppet, & Tulumello, 2000).

Other negotiators view such narrowing strategies as leading to lowest common denominator agreements that limit the value of agreements and leave each party only a little better off in comparison to what they stand to gain by acting on their own. According to this view, negotiators should aim, instead, to "enlarge the mythical fixed-pie" (Bazerman & Moore, 2012), that is design more ambitious agreements that build on the productive elements of conflict to produce more benefits for each participant. One strategy to do so is to integrate differences among the parties' interests (Lax & Sebenius, 1991; Lewicki, Saunders, & Barry, 2015; Mnookin et al., 2000; Susskind, 2014). Consider, for example, two children arguing over an orange. One wants it to bake a cake and the other to make juice. Instead of cutting the orange in half, they separate the peel and the pulp, each receiving greater value (R. Fisher, Ury, & Patton, 2011).

This chapter analyzes these three conflict management strategies across three cases of water diplomacy during periods of high tensions and rapid change: negotiations between 1856 and 1914 to establish the Danube navigation regime, negotiations beginning in the 1990s to create the Nile benefit sharing regime, and negotiations beginning in the 1980s to form the Danube water protection regime. Despite their unique morphologies and histories, the Danube and Nile River basins are two of only four international river basins that drain territory from ten or more countries. The cases therefore share similar challenges for organizing international cooperation. The negotiations also share a treaty-based format typical of water diplomacy. Comparing very different cases from the Global North and the Global South, each with its own distinct historical period, institutional context, problem focus, and conflict management strategies, increases the relevance of the analysis for environmental conflict management, in general, and for negotiators seeking to manage diverse water conflicts.[1] Additionally, each water diplomacy case involved significant conflicts over access to water resources and benefits from its use, as well as over access to participate in water decision-making, and therefore provides insight into access and conflict, two of this book's main themes. The analysis is based on over ten years of qualitative data gathered in the Danube and Nile basins, which include questionnaires, in-depth interviews, participant observation, and document analysis.

The Danube and Nile river systems

The Danube forms a single main river flowing 2857 kilometers west to east that drains a total area of 801,463 square kilometers with territories from nineteen countries and Kosovo (see Figure 5.1). From rushing headwaters and sparkling, yet dangerous, eddies, the Danube River meanders eastward across great plains, cuts its way through mountainous gorges, and finally reaches its delta, where it slowly flows into the Black Sea. Its watershed is home to 83 million people and rich biodiversity, including endemic species found nowhere else, with great ecological, cultural, and economic value for tourism and the fishing industry. Other vital uses include drinking water supply, water for agriculture, wastewater disposal, hydropower generation, and navigation.

The Nile forms two main branches south of Khartoum, the White Nile and Blue Nile (Abay), flows 6695 kilometers, and drains territories from about 3.2 million square kilometers with territories from eleven countries (see Figure 5.2). Of the 238 million people who live in the basin, many are vulnerable to water and food insecurity. The White Nile system includes Lake Victoria, the world's third largest freshwater lake by surface area, and the Sudd, a Ramsar wetland of international importance. During its rainy season, the Blue Nile's flow increases by up to twenty times and rushes downstream through a steep gorge before reservoirs, dams, and cataracts slow its flow through an increasingly arid and level plain. Downstream of Lake Nasser (Lake Nubia) and the Aswan High Dam in Egypt, the Nile becomes a regulated, year-round irrigation system that discharges into the Mediterranean Sea. Wet, mountainous areas contrast

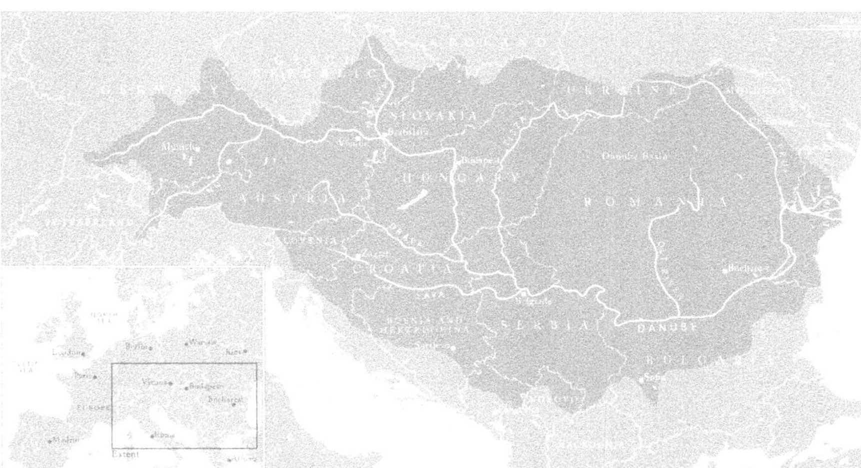

Figure 5.1 The Danube River Basin

Source: Cartography by Gregory Woolston using data from ICPDR retrieved from www.icpdr.org (accessed 3/1/2016), and Natural Earth.

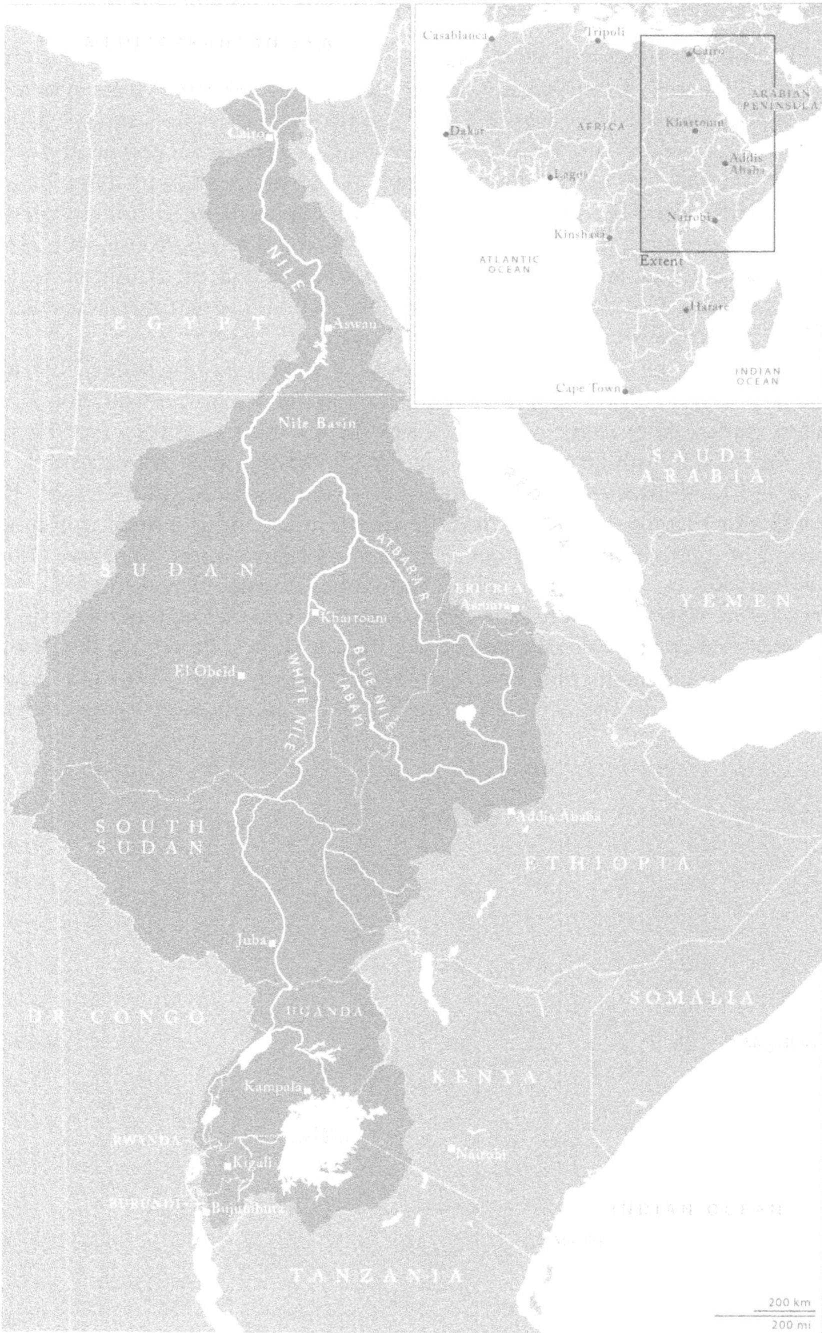

Figure 5.2 The Nile River Basin

Source: Cartography by Gregory Woolston using data retrieved from NBI, sob.nilebasin.org (accessed 3/1/2016), and Natural Earth.

starkly with arid downstream regions, creating stunningly diverse habitats for biodiversity and many endemic species important for tourism and fisheries. Other important water uses include hydropower generation, agriculture, wastewater disposal, local navigation, and shipping between Egypt and Sudan.

The three cases

The Danube navigation regime (1856–1914)[2]

The 1856 Treaty of Paris ended the Crimean War and created the first multilateral arrangements for the Danube. The riparian countries with territory in the basin (Austria and the Ottoman Empire) and the non-riparian countries without territory in the basin (France, Great Britain, Sardinia, Prussia, and Russia) disagreed over many issues. (Hoping to regain delta territory, Russia often sided with the riparians.) The negotiators created two commissions, the European Danube Commission (EDC) for governing the delta region, in which riparians and non-riparians participated, and the Riparian Commission (RC) for governing the entire river system. Only riparians participated in the short-lived RC, and, by 1857, they returned control of most river activities to individual riparians and restricted non-riparian ships' access between river ports, permitting them passage and commerce only between ports on the Danube and the open sea.

The Nile benefit sharing regime (1997–the present)

In the late 1990s, an alliance of upper riparian countries (Burundi, Rwanda, Tanzania, Uganda, Kenya, the Democratic Republic of the Congo (then Zaire), Ethiopia, and Eritrea), agreed to negotiate with downstream Egypt and Sudan toward a cooperative framework agreement (CFA) and transitional arrangements, the Nile Basin Initiative (NBI), to promote sharing a broad range of benefits from water access and use. A draft CFA was completed in 2007, but negotiations stalled. In 2010, over downstream objections, the upstream countries postponed resolution of the remaining contentious text until the formation of the Nile River Basin Commission (NRBC) and opened the CFA for signature. Egypt and Sudan subsequently stopped participating fully in the NBI (Sudan has since reactivated its membership) (Sewilam, 2015). The CFA will enter into force upon ratification by six signatories. Ethiopia, Rwanda, Tanzania, Uganda, Kenya, and Burundi have already signed, and South Sudan has indicated its intention to join.

The Danube water protection regime (1985–present)

During the Cold War, Danube negotiators initiated water protection discussions and, in 1985, two western countries, Germany and Austria, and six Central and Eastern European Countries (CEECs)—Bulgaria, Czechoslovakia, Hungary, Romania, the Soviet Union, and Yugoslavia—signed the Bucharest

Declaration to promote cooperation on water quality and quantity issues. Soon after, parallel negotiations began toward two treaties linking water and the environment: the Danube Basin Ecological Convention (also the Danube Basin Nature Conservation Convention), which ended without agreement, and the Convention on Co-operation for the Protection and Sustainable Use of the River Danube (also Danube River Protection Convention or DRPC), signed in 1994 by the eleven major Danube states, Austria, Bulgaria, Croatia, the Czech Republic, Germany, Hungary, Moldova, Romania, Slovakia, Slovenia and Ukraine, and the European Community. The DRPC created the International Commission for the Protection of the Danube River (ICPDR) and participants now also include Bosnia-Herzegovina, Montenegro, and Serbia. In 2000, the DRPC parties agreed to use the ICPDR to coordinate implementation of the European Union Water Framework Directive (EU WFD).

Evidence from the cases

Building on noncompetitive similarities

The following Danube and Nile cases provide very different examples of negotiators crafting agreements by narrowing the range of issues they addressed to those on which they agreed and excluding their differences in other areas. These strategies were effective for initiating negotiations, creating provisional arrangements, and realizing specific benefits of water use.

The Danube navigation regime

Both riparians and non-riparians wanted their ships to be able to access the Danube delta to engage in commerce, but it was not clear who should provide the physical and institutional conditions to ensure access. Previously, the Ottoman Empire maintained navigable conditions in the delta, but delta conditions had become dangerous under Russian control between 1829 and 1856. Sir Charles Hatley, a chief engineer of the delta's EDC, described the state of the delta:

> Shipwrecks were of common occurrence, and occasionally the number of disasters was appalling. One dark winter night in 1866, during a terrific gale, 24 sailing ships and 60 lighters went ashore off the [Sulina] mouth and upwards of 300 persons perished.
>
> (quoted in Chisholm, 1910: p. 822)

In 1856, the Ottoman Empire regained its delta territory, but lacked the financial resources to maintain delta access (Jensen & Rosegger, 1978).

Austria wanted to protect delta access for its shipping interests, which transported significant quantities of manufactured goods to downstream ports and the Black Sea. Delta access was also a priority for Great Britain, which exported wheat shipments from the lower Danubian principalities

(Chamberlain, 1923; East, 1932; Jensen & Rosegger, 1978). Russia, no longer a riparian, also wanted to ensure delta access. The countries built on their shared interest in ensuring access to the Danube delta and created provisional EDC arrangements for governing the delta region.

The Nile basin benefit sharing regime

The NBI aims to further "sustainable socio-economic development through equitable utilization of, and benefit from, the common Nile Basin water resources" (Nile Basin Initiative, 2013:3), help countries build confidence, and gain much-needed experience cooperating with one another. Uganda's Commissioner for Water Resources Planning and Regulation, an NBI participant, described the situation, "At the beginning, we would be in a room and we wouldn't talk to each other—we saw each other as enemies. No one would talk in meetings because of suspicion" (cited in Nile Basin Initiative, 2013).

The benefit-sharing approach also shifted the focus of negotiations away from bargaining over how specific water quantities will be distributed and to negotiating over noncompetitive, win–win benefits (Interview Nile 14, 2007) from sharing the broader set of ecosystem services provided by access to water (Phillips, Daoudy, McCaffrey, Öjendal, & Turton, 2006; Sadoff & Grey, 2002). For example, the NBI conducted a basin-wide study of regional power trade, which identified (1) high regional demand for electricity for domestic, industrial, and commercial purposes, and (2) diverse power options in the basin, including solar, wind, coal, geothermal, and significant hydropower potential. The NBI then facilitated implementation of regional investment projects to connect countries' national electric grids. Ethiopia now exports surplus hydropower power to Sudan, which Sudan uses to replace electricity sources that have higher greenhouse gas emissions. During low rainfall periods, Ethiopia can rely on Sudan's thermal production to protect its energy security. In this example, the countries built on shared objectives to reduce costs from generating electricity and benefit from excess capacity (Nile Basin Initative, 2014a).

The Danube water protection regime

Due to Cold War tensions, at the start of water protection negotiations the Danube countries had little experience negotiating with one another and no information about environmental issues across the basin, but the 1980s and early 1990s brought significant political changes, catalyzing a shared interest in addressing environmental and security risks. When Romania and Yugoslavia first contacted the upstream countries to discuss water management and political issues, water protection was less of a priority for upstream countries, but they were generally interested in collaborating in order to learn about the environment and politics in the Eastern Bloc (Interview Danube 17, 2006). At the same time, Ukraine's Chernobyl nuclear accident, accidental pollution of the Rhine from the Swiss Sandoz agrochemical warehouse, and algal blooms in the Black Sea near the mouth of the Danube provided visible evidence of transboundary risks. The countries built on their noncompetitive interest and initiated negotiations to reduce both transboundary environmental and security risks.

Integrating differences

In contrast, the following examples analyze conflict management strategies intended to build on the productive aspects of conflict with the goal of producing greater benefits, in comparison to what negotiators could achieve by focusing only on noncompetitive similarities. Negotiators identified issues over which they disagreed and integrated their preferences across four kinds of differences: sequencing, relative valuations, forecasts and risk preferences for the future.

Integrating sequencing preferences

Negotiators can sequence elements of their agreements in order to manage their different preferences for which issues to address first and which later, as well as differences over the timing of implementing new arrangements (Lax & Sebenius, 1991). In the Nile and Danube cases, specific sequencing strategies included creating provisional arrangements, splitting negotiations into parallel tracks for short- and long-term arrangements, and creating a framework agreement with easily revisable appendices. However, these strategies were effective only when they included clear timetables to demonstrate progress and benchmarks against which to evaluate it.

THE DANUBE NAVIGATION REGIME

At the end of the Crimean War, the riparians and non-riparians had no experience cooperating with one another on navigation, and neither coalition had confidence the other would protect its interests. The non-riparians prioritized an arrangement providing urgent delta access for their ships. The riparians shared this interest but prioritized long-term arrangements for governing the entire navigable river. The parties integrated these differences in a sequenced agreement. A provisional, two-year agreement ensured access to the delta for all ships and gave the riparians time to develop rules for governing the entire Danube and all countries an opportunity to gain experience cooperating with each other. At the end of the two years, a long-term arrangement could replace the provisional arrangements.

THE NILE BASIN BENEFIT SHARING REGIME

The Nile negotiators disagreed over which issues they should address first. The downstream countries wanted to start cooperating on non-competitive similarities such as technical studies, but upstream countries were concerned negotiating only these issues would not address their priority, changing the status quo of water use across the basin (Interview Nile 03, 2004; Interview Nile 14, 2007). Ethiopia, therefore, insisted on creating legally binding arrangements for water utilization as a first step, which they also knew might take a long time (Arsano & Tamrat, 2005).

The Nile countries managed their different sequencing preferences by initiating two parallel negotiation tracks. In 1997, negotiations began on the CFA and, in 1999, on the NBI, which was intended as a "transitional institutional mechanism…pending the conclusion of a Cooperative Framework Agreement" (*The Nile Basin Initiative Act*, 2002). As one Ethiopian negotiator explained:

> We need to proceed on the two-track process…It might take time to resolve those [CFA] issues. But, we need to also see the actual benefits that might come out of cooperation. So, let's proceed with the current project and also…try to see the different models of how we can negotiate and maybe that will give us a breakthrough.
>
> (Interview Nile 09, 2004)

The provisional NBI was expected to build negotiating capacity and trust, which would facilitate negotiations on the more contentious, long-term CFA (Interview Nile 19, 2007).

However, the CFA's slow progress and the NBI's lack of demonstrable benefits led to upstream concerns that the provisional arrangements were a downstream strategy to delay changes in upstream water use and access. The NBI's arrangements did not define short-, medium-, and long-term timeframes for realizing urgently needed benefits (Cascão, 2013; Nile Basin Initiative, 2014b). An Ethiopian participant expressed this frustration: "People are engaged in trying to identify win-win projects, but so far there have been no big projects, which is causing some shortcomings in confidence" (Interview Nile 19, 2007). Additionally, the sequencing strategy did not specify a timeframe for completing the CFA negotiations, or what would happen if the negotiations failed. By 2007, the CFA negotiators agreed on all but Article 14 on water security. The upstream countries proposed another sequencing strategy: sign the CFA and postpone resolution of Article 14(b) until establishment of the NRBC. However, Egypt and Sudan rejected this suggestion, insisting that agreement on the article's text come first (Mekonnen, 2010). Although the sequencing arrangements enabled cooperation to develop, without a timetable they were insufficient, and the future of the NBI and CFA is now uncertain.

THE DANUBE WATER PROTECTION REGIME

In the 1980s, when Romania invited the seven other major Danube states to negotiate about water quality and the environment, all countries were uncertain whether they would be able to reach agreement. Even if they could reach agreement, without trust or access to information about compliance, they knew they faced significant implementation obstacles (Interview Danube 16, 2006).

The negotiators disagreed over the form of the agreement: the central and eastern countries wanted a binding treaty, while upstream Germany and Austria preferred nonbinding recommendations. The negotiators reached agreement on a provisional, nonbinding declaration (Hock & Kovacs, 1987). Germany and Austria viewed the declaration as creating an important moral and political obligation to continue cooperating (Interview Danube 01, 2006; Interview Danube 04, 2006), allowing the countries to "…first accomplish something and only later address other issues" (Interview Danube 16, 2006). The CEECs considered the agreement acceptable because, according to a Romanian participant, they saw it as the first step in a process that could lead to a future international convention (Varduca, 1997). It was "…just a beginning" (Interview Danube 20, 2006), and it could "serve as a kind of model for a future treaty" (Interview Danube 17, 2006).

In the early 1990s, all DRPC negotiators wanted to reach a binding agreement within a short period of time but recognized they didn't know enough about complex transboundary pollution issues and what it meant to address problems at the basin scale. They therefore applied a sequenced approach, agreeing to sign the convention and include technical appendices, which could easily be revised without subsequent ratification as information became available (Interview Danube 03, 2006; Interview Danube 04, 2006). The sequenced convention and appendix strategy enabled them to start "early but imperfectly" (Jansky, 2002:66), evaluate progress, and later revise. For example, negotiators included water quality standards in the initial agreement's appendix even though they could not yet implement harmonized processes for data collection and sharing. The negotiators also anticipated this flexible design would help the DRPC meet future EU WFD requirements (Interview Danube 03, 2006). Today, the EU WFD's timetable for implementation and signposts for evaluating progress provide mechanisms for ongoing adjustment of the DRPC.

Integrating different relative valuations

Even when the parties value all issues being negotiated, they may prioritize some issues over others. Negotiators can trade across these different priorities, gaining more benefits on their higher priority issues, while accommodating others' interests on issues they care less about (Mnookin et al., 2000). In the following cases negotiators effectively built on their different priorities for the geographic scope of cooperation, the scope of activities and problems they would address, the authority invested in the international implementing organization, the form of the agreement, and access to the benefits of using water.

THE DANUBE NAVIGATION REGIME

During the navigation negotiations the non-riparians' top priority was to protect their shipping interests by ensuring passage for ships from all countries along the entire navigable length of the Danube and universal access to all ports for commerce. The British, in particular, wanted to increase grain exports from the lower Danubian principalities and prevent Austria from acquiring a monopoly over Danube commerce. Riparians wanted to protect their shipping interests by limiting non-riparian access upstream of the delta and along the river (Kaeckenbeek, 1962; Krehbiel, 1918; Sherman, 1923).

The riparians' top priority was to protect their ability to make decisions within their territories by excluding non-riparians from Danube navigation administration and limiting the authority of any international organization. Riparians especially wanted to retain control over implementation of river maintenance and improvement projects as well as the right to levy fees to pay for these projects. To protect their shipping interests, non-riparians wanted to participate in an international authority with expansive powers. France pointed to Russia's failure to maintain the delta as justification for universal participation in, at least, regulating and implementing delta maintenance (Kaeckenbeek, 1962).

The Danube countries created two separate arrangements, which together integrated the riparians' and non-riparians' different relative valuations. The EDC included participation from both riparians and non-riparians and was given two-year provisional authority to implement maintenance and improvement projects in the delta. The RC's membership was restricted to riparians and it was given authority to develop navigation rules for the entire Danube and coordinate, but not implement, navigation improvement projects. After two years, the RC was to assume the authority of the EDC (Donaukommission, 2004; Gourdon, 1857; Kaeckenbeek, 1962; Krehbiel, 1918). The riparians accepted short-term internationalization of the delta under a powerful authority and permitted passage for all ships along the entire navigable Danube in exchange for long-term, limited riparian control over the entire river system. The non-riparians accepted the arrangements because their interests in delta access and river passage were protected.

THE DANUBE WATER PROTECTION REGIME

In the Bucharest Declaration negotiations the main issues were the legal form, the scope of problems and activities, and the geographic scope. Germany and Austria's top priority was the form of the agreement and both wanted to negotiate only nonbinding recommendations. According to a German participant, "We had almost no opportunity to verify what was happening in the CEECs in practice…For us, it would have been unthinkable to bind ourselves one-sidedly" (Interview Danube 17, 2006). Germany was unwilling to commit to further action to improve water quality because it could not ensure the central and eastern countries would implement their commitments. Most CEECs wanted a binding agreement, but this was not their top priority. Many CEEC negotiators' priority interest was to address a broad scope of activities, including preventing harm to downstream countries as a result of low flows, ecological protection, monitoring, and data exchange (Varduca, 1997). The resulting Declaration is nonbinding and covers a broad scope of water quantity and quality activities. Germany and Austria would have preferred a narrower agreement, but, from their perspective, they avoided incurring nonreciprocal obligations. The CEECs would have preferred a legally binding agreement, but they could agree because of the Declaration's broad scope of activities.

When all countries agreed to negotiate a legally binding agreement, they still disagreed about what scope of activities and problems to address. Hungary proposed a comprehensive agreement on a broad range of ecological issues, including sustainable resource use and the preservation and restoration of biodiversity as well as landscape diversity, especially wetlands (Temporary Secretariat Danube Basin Ecological Convention, 1998). Germany and Austria disagreed, preferring a narrower focus on protecting water and the riparian environment from transboundary pollution, which they thought was a more practical strategy to reach agreement within a short timeframe (Interview Danube 03, 2006). A Czech delegate recalled that most former CEECs wanted to include the issue of water quantity, but Germany and Austria disagreed (Interview Danube 07, 2006).

The DRPC negotiators also disagreed about whether the agreement should apply to the entire basin and how to define transboundary impacts. Austria and Germany wanted to ensure all countries incurred reciprocal obligations, and, with Czechoslovakia, wanted to include the entire basin, but Romania disagreed, preferring to include only some parts of the basin. Austria was concerned Romania would exclude its mining wastewater ponds and not control pollution from them. Austria also wanted to define transboundary impacts to ensure that all countries would be obliged to control all pollution exceeding their territories, including the flow of pollution from their territories directly into the Black Sea (Interview Danube 03, 2006). Hungary was still concerned the Austrian approach focused too much on the main river and not enough on the basin environment.

The DRPC integrated negotiators' different relative valuations; it is legally binding and covers the entire basin and all adverse transboundary impacts but addresses a narrow range of water quality issues. Germany and Austria agreed because the scope and definition of transboundary impacts ensured reciprocal obligations over a limited range of issues. The other countries could agree because the Convention was legally binding and some accommodations were made to address their interest in a broader range of activities and problems. For example, the DRPC's name was changed from *Water Management* to *Protection and Sustainable Use*, and the agreement includes requirements for wetlands. Even though Austria and Germany were unwilling to include an even stronger ecological focus at the time, all negotiators expected the pending EU WFD would require them to address ecological aspects of water protection in the future (Interview Danube 03, 2006; Interview Danube 25, 2006; Interview Danube 33, 2006).

Integrating differences in forecasts and risk preferences

Negotiators may also hold different expectations about future events such as the extent of a dam's flooding impacts, and different risk preferences such as how tolerant a party is of risk that the dam will cause negative downstream impacts. In a contingent agreement, negotiators avoided arguing about uncertain future events by deferring finalization of details until the contingency occurs, essentially adopting a "wait and see approach" (Bazerman & Gillespie, 1999). For example, negotiators could monitor actual flooding impacts after dam construction and then revise the dam's operation according to a previously agreed upon strategy. As the following cases show, in order to be effective contingent agreements need to identify the issues for which negotiators have the lowest risk tolerance, minimize the appropriate risk, clearly identify the conditions triggering the agreed-upon contingency, and specify strategies for implementing and enforcing agreed-upon alternatives.

THE DANUBE NAVIGATION REGIME

The EDC was designed as a two-year arrangement for the delta region during which time the riparians would develop long-term rules for the entire Danube. However, the EDC could only be dissolved by unanimous consent of its seven riparian and non-riparian members. The non-riparians did not trust the riparians to enact long-term rules protecting their shipping interests and were especially risk-averse when it came to the possibility of being excluded from commercial access to ports. In the event the riparians enacted unfavorable rules, the unanimous decision rule meant non-riparians could refuse to dissolve the EDC and protect delta access for their ships. The riparians could not be sure the non-riparians would agree to dissolve the EDC, but the contingent arrangement limited other countries' control over the upper and middle Danube, protecting their priority interest. Instead of arguing over each other's intentions, the contingent rules minimized risk to both coalitions' priority interests.

THE NILE BASIN BENEFIT SHARING REGIME

So far, the upstream and downstream countries in the Nile basin have not been able to successfully address their differing expectations for the CFA and varied tolerances for possible future negative impacts. By signing the CFA most upstream countries indicated they expect the CFA will protect their interests, while Egypt and Sudan do not have this confidence and have refused to sign. The most contentious issue is Article 14 on water security. Negotiating as a united coalition for the first time in history (Cascão, 2013), all seven upstream countries agreed on text for Article 14(b), "…not to significantly affect the water security of any other Nile Basin States." Sudan and Egypt disagreed, believing this could imply consent to future upstream developments with unknown potential harm to existing downstream uses. Egypt proposed substituting "… not to adversely affect the water security and current uses and rights of any other Nile Basin State" (Agreement on the Nile River Basin Cooperative Framework, 2010), which is evidence of the downstream countries' low tolerance for the possibility of negative impacts. Whether or not upstream countries expect future development to have negative downstream impacts, they have a higher tolerance for this possibility. The upstream countries rejected Egypt's proposal, reflecting their low tolerance for the possibility that downstream countries could use it to veto future upstream water development.

Article 5 of the CFA on the obligation not to cause significant harm includes a contingent approach but lacks the implementation details needed to effectively manage differing expectations and risk preferences (Bazerman & Gillespie, 1999). Article 5 requires a state causing significant harm to another basin state to eliminate or mitigate the harm, or to provide compensation. However, it does not specify a threshold for unacceptable transboundary impacts that would trigger the need for action, a strategy for determining the appropriate response, or how these actions would be enforced. The CFA instead defers development of specific implementation measures to the future NRBC. Interestingly, Egypt, Sudan, and Ethiopia agreed to an almost identical approach in the 2015 Agreement on Declaration of Principles on the Grand Ethiopian Renaissance Dam (GERDP), in which they defer specifying the implementation measures for a contingent agreement in the event of harm (Official Text: Egypt, Ethiopia, Sudan – Declaration of Principles, 2015). Why did Egypt and Sudan sign the GERDP, but not the CFA, when both agreements include almost identical contingency clauses in the event of significant harm? A possible explanation, based on protecting dissenting interests, will be discussed later.

THE DANUBE WATER PROTECTION REGIME

Hungary and Austria held different expectations for the negotiators' ability to reach agreement on a broadly focused ecological convention. They managed these differing expectations by deciding to initiate two negotiation tracks for separate Danube conventions. Hungary led negotiations on the broadly focused Danube Basin Ecological Convention, and Austria led negotiations on the more narrowly focused DRPC. The countries avoided arguing over what was possible by pursuing both options and seeing what happened.

Party arithmetic

In order to reach agreement, it may seem expedient to exclude parties with dissenting interests from participating in water policy decisions and dispute resolution procedures. However, as these cases demonstrate, this narrowing strategy, intended to minimize the negative aspects of conflict, can instead exacerbate conflict and encourage excluded parties to retaliate and undermine collaborative efforts. Strategies negotiators can use to protect dissenting interests include consensus decision rules, giving vulnerable countries weighted votes, and creating forums for discussing differences (Posner & Sykes, 2014).

The Danube navigation regime

The EDC members made decisions by majority vote. Their decisions did not require subsequent ratification and were binding on all members, regardless of how they voted. Although at first most of the EDC members were not riparian to the delta, the year 1878 brought many geopolitical changes: the Ottoman Empire lost its riparian status and new delta region riparians emerged—Romania, Serbia, an autonomous Bulgaria, and Russia. Romania was admitted to the EDC, but Serbia and Bulgaria were not (Russia was already a member). The EDC's decision rule enabled the coalition of non-delta riparians to exclude the new delta riparians' interests and act contrary to them, for example by expanding the EDC's geographic scope and authority against their wishes. For example, when the EDC decided to create a Mixed Commission with authority over part of the Danube flowing through Romania, Bulgaria, and Serbia, the majority decided to invite Serbia and Romania to participate in negotiations in a consultative, not voting, capacity. Romania and Bulgaria protested what they saw as an infringement of their sovereignty and refused to participate, but the majority voted against their interests. Romania subsequently stopped participating in the EDC and actively worked to undermine its authority and render it ineffective (Kaeckenbeek, 1962; Sherman, 1923; Teclaff, 1991).

The Nile benefit sharing regime

Egypt and Sudan wanted to come to an agreement on the CFA's Article 14(b) on water security before opening the agreement for signature, but the upstream majority outvoted them, opened the CFA for signature, and deferred resolution of Article 14(b) until the establishment of the NRBC. One explanation for Egypt and Sudan's refusal to sign the CFA is their lack of confidence its decision-making, dispute resolution, and amendment procedures will protect their interests from the upstream majority's future actions.

Although many of the NRBC's rules and procedures are deferred for later decision, the CFA specifies that if its governing body, the Council of Ministers (COM), cannot reach consensus on a decision, such as the text of Article 14(b), a fact-finding commission (FFC) will resolve the dispute. The FFC would be comprised of one member from each state concerned, which in this case would probably include all Nile member states, plus one member of a different nationality. The FFC would make decisions by majority vote, which could enable an upstream majority to outvote Egypt and Sudan. During the negotiations, the upstream and downstream countries also disagreed over the CFA's amendment procedure (Salman, 2013). Although many CFA elements, including Article 14, can be amended only by consensus, the CFA's dispute resolution procedures can be amended by a two-thirds supermajority vote, which means that downstream interests could be outvoted in the event of a dispute.

Taken together, these CFA procedures provide insight into the question posed earlier, "Why did Egypt and Sudan sign the GERDP, but not the CFA?" GERDP includes participation from only Egypt, Sudan, and Ethiopia. The two downstream riparians can therefore protect their interests in any majority vote. In contrast, the CFA is open to participation from all Nile Basin countries and there are more upstream riparians than downstream riparians in the basin.

According to the CFA's provisions, the upstream riparians could act contrary to downstream interests.

The Danube water protection regime

During the DRPC negotiations, the countries required unanimous agreement on all elements, which provided confidence that all participants' environmental interests would be protected. However, from the start, the water protection negotiations deliberately excluded navigation interests, with western diplomats intentionally deciding against expanding the existing, Soviet-dominated Navigation Commission's mandate (Interview Danube 01, 2006). Creating separate spaces for negotiating navigation and environmental decisions has led to contemporary conflicts between these two interests. For example, as part of a European effort to better integrate national transport networks, shipping interests would like to increase the tonnage of goods transported on the Danube (European Commission, 2009). Removing the most significant barriers would require deepening some Danube segments, including the main navigation bottleneck, a 69-kilometers segment between Straubing and Vilshofen, Germany. Environmentalists view this segment, one of the only "free flowing" parts of the upper Danube not already impacted by locks, dams, and canals, as ecologically invaluable, and they mobilized opposition.

In 2007, responding to the need for a forum to manage such conflict, the countries initiated the "Joint Statement on Guiding Principles on the Development of Inland Navigation and Environmental Sustainability in the Danube River Basin" and, with it, a process for information exchange and joint studies. By creating space for negotiators to discuss and integrate differences between environmental and navigation interests, this strategy is already managing conflict more productively (Interview Danube 15, 2006).

Discussion

In the diverse cases analyzed in this chapter, no single conflict management strategy determined negotiators' success. In all three, focusing on issues on which negotiators agreed allowed participants to gain experience cooperating with one another. However, building on noncompetitive similarities created a basis for longer-term agreements only when complemented with other strategies that took advantage of conflict's productive potential, such as sequencing arrangements, integrating negotiators' different preferences, and using contingent agreements with clear implementation mechanisms to manage their different risk preferences and expectations for the future. The Danube and Nile histories also offer insight into the risks of avoiding conflict through party arithmetic. These cases show that dissenting interests should be able to participate in water policy decisions and be protected if for no other reasons than to avoid encouraging excluded parties to undermine collaborative efforts and to avoid retaliation when today's dissenting interests become part of a future majority.

Based on these findings, future research on conflict management should focus on the interplay between strategies by which negotiators minimize the negative aspects of conflict (without excluding dissenting interests), while also building on conflict's productive aspects. Together, as part of a comprehensive package, these strategies create opportunities for valuable interaction and dialogue in river spaces that are subject to inherent conflict.

Acknowledgements

This research was funded in part by the Program on Negotiation at the Harvard Law School, the MIT Martin Family Society of Fellows for Sustainability, and Resources for the Future. I thank Dr. Alexis Peri and Dr. Christina Rosan for helpful comments on an earlier draft.

Notes

1 Even though two cases are located in the Danube River system, they have completely distinct negotiating contexts and institutions governing behavior.

2 Today, the Danube countries cooperate on navigation issues under distinct institutional arrangements created by the 1948 Convention Regarding the Regime of Navigation on the Danube and its amendments.

Works cited

Agreement on the Nile River Basin Cooperative Framework. (2010).

Arsano, Y., & Tamrat, I. (2005). Ethiopia and the Eastern Nile Basin. *Aquatic Sciences*, *67*, 15–27.

Bazerman, M. H., & Gillespie, J. J. (1999). Betting on the future: The value of contingent contracts. *Harvard Business Review*, September–October, 155–160.

Bazerman, M.H., & Moore, D.A. (2012). *Judgment in managerial decision making (8th Edition)*. New York: John Wiley & Sons.

Cascão, A. E. (2013). Nile water governance. In S. B. Awulachew, V. Smakhtin, D. Molden, & D. Peden (Eds.), *The Nile River Basin: Water, agriculture, governance and livelihoods* (pp. 229–252). Abingdon, UK: Routledge.

Chamberlain, J. P. (1923). *The regime of the international rivers: Danube and Rhine*. New York: Longmans, Green & Co.

Chisholm, H. (1910). Danube. In *The Encyclopaedia Britannica: A dictionary of arts, sciences, literature and general information (11th Edition)* (p. 822). New York: The Encyclopedia Brittanica Co.

Donaukommission. (2004). *Die Donaukommission und die Donauschifffahrt*. Budapest: Donaukommission.

East, W. G. (1932). The Danube route-way in history. *Economica*, *37*, 321–345.

European Commission. (2009). TEN-T guidelines. Retrieved from http://ec.europa.eu/transport/infrastructure/basis_networks/guidelines/guidelines_en.htm

Fisher, J. (2014). Managing environmental conflict. In P. T. Coleman, M. Deutsch, & E. C. Marcus (Eds.), *The handbook of conflict resolution* (Third ed.). San Francisco, CA: John Wiley & Sons, Inc.

Fisher, R., Ury, W., & Patton, B. (2011). *Getting to yes: Negotiating agreement without giving in*. New York: Penguin Books.

Gourdon, E. (1857). *Histoire de Congrès de Paris*. Paris: Librairie Nouvelle.

Hock, B., & Kovacs, G. (1987). Annex I (Draft translation from Hungarian) Declaration on the Cooperation of the Danube Countries on Water Management and Especially Water Pollution Control Issues of the River Danube. In B. Hock & G. Kovacs (Eds.), *A Large International River (WP-87–11)*. Vienna: IIASA.

Interview Danube 01. (2006). Vienna, Austria.

Interview Danube 03. (2006). Vienna, Austria.

Interview Danube 04. (2006). Vienna, Austria.

Interview Danube 07. (2006).

Interview Danube 15. (2006).

Interview Danube 16. (2006). Munich, Germany.

Interview Danube 17. (2006). Munich, Germany.

Interview Danube 20. (2006). Bucharest, Romania.

Interview Danube 25. (2006). Vienna, Austria.

Interview Danube 33. (2006). Vienna, Austria.

Interview Nile 03. (2004). Addis Ababa, Ethiopia.

Interview Nile 09. (2004). Addis Ababa, Ethiopia.

Interview Nile 14. (2007). Nairobi, Kenya.

Interview Nile 19. (2007). Addis Ababa, Ethiopia.

Jansky, L. (2002). The river Danube: Needs for integrated river management. In L. Jansky, M. Nakayama, & J. Uitto (Eds.), *Lakes and reservoirs as international water systems: Towards world lake vision* (pp. 55–68). Tokyo: UNU Press.

Jensen, J. H., & Rosegger, G. (1978). Transferring technology to a peripheral economy: The case of Lower Danube transport development, 1856–1928. *Technology and Culture*, *19*, 4, 675–702.

Kaeckenbeek, G. (1962). *International rivers: A monograph based on diplomatic documents*. New York: Oceana Publications Inc.

Krehbiel, E. (1918). The European Commission of the Danube: An experiment in international administration. *Political Science Quarterly*, *33*, 1, 38–55.

Lax, D. A., & Sebenius, J. K. (1991). Thinking coalitionally: Party arithmetic, process opportunism, and strategic sequencing. In H. P. Young (Ed.), *Negotiation analysis* (pp. 153–193). Ann Arbor, MI: The University of Michigan Press.

Lewicki, R. J., Saunders, D. M., & Barry, B. (2015). *Negotiation* (Seventh ed.). New York: McGraw-Hill Education.

Mekonnen, D. Z. (2010). The Nile Basin Cooperative Framework Agreement negotiations and the adoption of a "water security" paradigm: Flight into obscurity or a logical cul-de-sac? *The European Journal of International Law*, *21*, 2, 421–440.

Mnookin, R. H., Peppet, S. R., & Tulumello, A. S. (2000). *Beyond winning: Negotiating to create value in deals and disputes*. Cambridge, MA: Belknap Press.

Nile Basin Initiative. (2013). *Cooperation on the Nile*. Entebbe, Uganda: Nile Basin Iniative.

Nile Basin Initiative. (2014a). *Harnessing the Nile to increase access to electricity: The Ethiopia-Sudan transmission interconnection project*. Entebbe, Uganda: Nile Basin Iniative.

Nile Basin Initiative. (2014b). *Nile cooperation: Opportunities & challenges*. Entebbe, Uganda: Nile Basin Initiative.

Official Text: Egypt, Ethiopia, Sudan – Declaration of Principles. (2015). *Horn Affairs English*. Retrieved from http://hornaffairs.com/en/2015/03/25/egypt-ethiopia-sudan-agreement-on-declaration-of-principles-full-text/

Phillips, D., Daoudy, M., McCaffrey, S., Öjendal, J., & Turton, A. (2006). *Trans-boundary water co-operation as a tool for conflict prevention and broader benefit sharing*. Stockholm: Ministry for Foreign Affairs, Sweden.

Posner, E., & Sykes, A. O. (2014). Voting rules in international organizations. University of Chicago Public Law and Legal Theory Working Paper No. 458.

Pruitt, D., Rubin, J., & Kim, S. H. (2003). *Social conflict: Escalation, stalemate, and settlement* (Third ed.). New York: McGraw-Hill Education.

Sadoff, C. W., & Grey, D. (2002). Beyond the river: The benefits of cooperation on international rivers. *Water Policy*, *4*, 389–403.

Salman, S. M. A. (2013). The Nile Basin Cooperative Framework Agreement: A peacefully unfolding African spring? *Water International*, *38*, 1, 17–29.

Sewilam, M. (2015). Egypt and the Nile Basin Initiative. Retrieved from http://weekly.ahram.org.eg/News/12818/17/Egypt-and-the-Nile-Basin-Initiative.aspx

Sherman, G. E. (1923). The international organization of the Danube under the peace treaties. *The American Journal of International Law*, *17*, 3, 438–459.

Susskind, L. (2014). *Good for you, great for me: Finding the trading-zone and winning at win-win negotiation*. Philadelphia, PA: Public Affairs.

Teclaff, L. A. (1991). Fiat or custom: The checkered development of international water law. *Natural Resources Journal*, *31*, 45–73.

Temporary Secretariat Danube Basin Ecological Convention. (1998). Danube Basin Nature Conservation Convention (Draft).

The Nile Basin Initiative Act (2002). Retrieved from http://faolex.fao.org/docs/pdf/uga80648.pdf

Varduca, A. (1997). The 1985 Bucharest Declaration: An important step toward Danube water quality protection. In I. L. Murphy (Ed.), *Protecting Danube River Basin resources* (pp. 31–41). Dordrecht: Kluwer.

6 The era of big dam building

It ain't over till it's over

Francis J. Magilligan, Chris S. Sneddon, and Coleen A. Fox

Introduction

Standing atop the McPherrin Dam on Butte Creek, California, in 1998, sledgehammer in hand, former Secretary of the Interior Bruce Babbitt declared the era of big dam construction to be over. He then made the first of many blows to the dam, initiating the new era of rethinking dams: "Dams are not like the pyramids of Egypt that stand for eternity. They are instruments that should be judged by the health of the rivers to which they belong" (Egan, 1998). Since then, approximately 1,100 dams have been removed nationally (American Rivers, 2014; O'Connor, Duda, & Grant, 2015), with hundreds more under consideration for removal for economic, environmental, or structural reasons. Despite a concerted effort to remove dams in the United States, and indeed in many other countries, large dam construction is poised to make an international comeback. Governments throughout the developing world are currently planning for and constructing thousands of dams in spite of local, regional, and international opposition efforts motivated by projected social and ecological consequences. The contemporary proliferation of dam building represents a global environmental conundrum for sustainable watershed management: what are the political, social, and environmental processes that, on one hand, question the continued efficacy of large dams and, on the other, push forward a global dam construction agenda?

Large dams have long been seen as a major engine of international development. With their supply of (allegedly) cheap hydroelectricity for domestic consumption and industrial expansion, their potential to expand irrigated agriculture, and their projected flood control benefits, large dams have long been envisioned as a fundamental resource for alleviating regional and global poverty and helping to modernize developing countries (Molle, 2009; White, 1957). Seen as drivers of economic growth, large multipurpose dams have also been an element of global geopolitical struggles during the Cold War (Agnew, 2011; Sneddon, 2015; Sneddon & Fox, 2011). For these combined economic, political, and strategic reasons, construction of large dams accelerated throughout the Global South, especially between the 1960s and 1980s (Figure 6.1), offset by roughly a decade

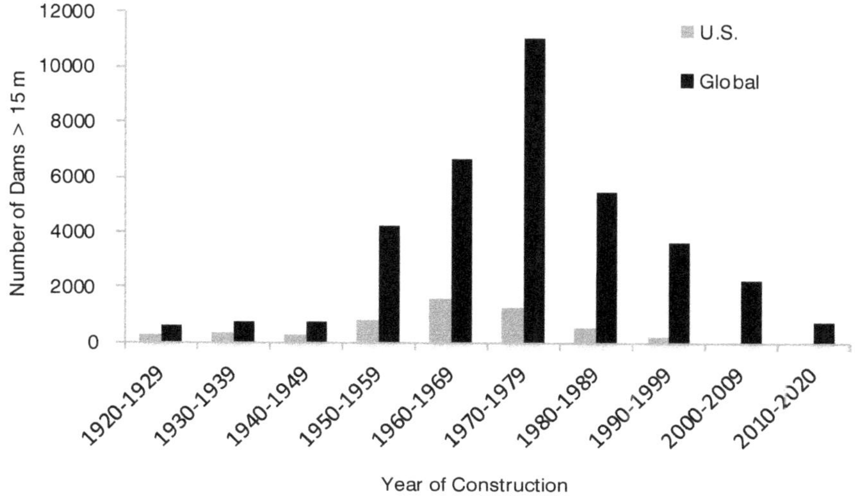

Figure 6.1 Construction of large (> 15 m) dams by decade for the United States and globally

Note: U.S. data removed from global totals.

Source: Data from the World Register of Dams.

from the intensification of large dam construction in the United States (Figure 6.1), which peaked during the 1960s.

Although large dam construction has attenuated in recent years, there is considerable concern that a new global era of large dam construction is slowly gaining momentum, despite the vast literature chronicling the profound social, economic, and environmental costs of damming rivers (cf. Ansar et al., 2014). This acceleration of large dam construction is in part due to China becoming a global economic powerhouse and subsequently mobilizing its economic and natural resources and technical expertise to construct numerous large dams domestically, especially along the upper reaches of the Mekong River (Dore et al., 2007). The Chinese government's domestic program is paralleled internationally as it eyes Africa, South America, and nearby Southeast Asia as sites of economic resources, geopolitical influence, and development.

This international spurt of large dam construction is not limited to China's global reach; many other countries, especially Iran and Turkey, are in the early planning stages of a robust increase in large dam construction both domestically and globally. At regional scales, environmental advocates have expressed dismay at rapid dam-building programs pushed by governments in the Balkans (Nelson, 2015) and the Himalayas (Ahlers et al., 2014; Grumbine & Pandit, 2013), with seemingly little concern for the social and ecological impacts of such projects. Moreover, the World Bank—which, in the face of intense pressure from environmental nongovernmental organizations (NGOs) and re-shifting economic priorities, had greatly curtailed its support of large dam construction by the 1990s—has recently announced that it is "back in the game" of large dam construction as a vehicle for reducing global poverty (Schneider, 2013; World Bank, 2009).

In this chapter, we focus on some of the recent quantitative trends in large dam construction in order to present a broad regional and global synthesis of ongoing trajectories of river alteration. The data come primarily from the World Register of Dams, the data set compiled by the ICOLD (International Commission on Large Dams, 2015). Where possible, we have augmented these data with data from the literature and other database surveys. Following our discussion of the distribution of large dams, we turn to a brief consideration of large dams' major biophysical impacts, focusing on the political drivers of dam construction and setting the stage for an in-depth analysis of one of the major questions framing contemporary debates about hydropower development: whether or not hydroelectricity can contribute to global efforts to mitigate climate change by lessening carbon emissions. We conclude with some thoughts on the future of damming in general, highlighting the need to rethink human interrelations with water in a way that de-emphasizes large dams as the primary means of "developing" rivers and using fresh water.

Our primary goal in this paper is to identify some of the key regional hotspots where controversial dams have been proposed and are under construction and to discuss some of the broader social, biophysical, and environmental dimensions of these trends. We argue that ongoing debates over the efficacy of large dam construction reveal important aspects of the contemporary politics of fresh water. Scientists, analysts, and activists hoping to intervene in these debates must be attentive to the scalar configurations—local, national, regional, and global—as well as the place-specific environmental consequences of large dams.

Distribution of dams worldwide

Determining the actual number of dams in the world is very difficult. Besides the problems of recording and reporting attributes and locations, the definition of what actually counts as a dam, and in particular as a "large" dam, varies. In some instances it refers to a particular operation or function, reservoir size, height, or other structural components (e.g. construction material or design). In other instances, the definition may simply refer to any barrier that impedes or restricts flow—such as farm ponds or check dams. A case in point: the U.S. Army Corps of Engineers maintains a national dams database, the National Inventory of Dams (NID), which defines a dam according to certain structural attributes, including height and volume. To be listed in the inventory, a dam must be (a) at least eight meters in height with a reservoir storage capacity of 18,500 cubic meters, or it must be (b) at least two meters in height with a reservoir storage capacity of 61,670 cubic meters. Based on these requirements, there are over 80,000 documented dams in the United States—in addition to at least 2.6 million small water bodies if one is referring to any artificial instream barrier (Smith et al., 2002).

Most dams meeting NID qualifications were constructed during the halcyon years of dam construction in the United States, 1950–1970 (Figure 6.1), when multiple state and federal agencies, with support from Congressional pork barrel politics, were involved with flood control, water supply, and energy provision (Billington & Jackson, 2006; Bromley, 2001; Gottlieb, 1988; Klein 1999; Molle, Mollinga, & Wester, 2009; Worster, 1985). At the time, the only major restriction limiting federally supported dam construction was a cost–benefit requirement of the 1936 Flood Control Act that simply stipulated that the economic benefits of a project need to exceed the costs. Although this requirement was initially designed to limit costly dam construction, in practice, it led to more extensive social and environmental impacts because it promoted larger, multipurpose project construction. The incremental costs of adding a new use often amounted to less than the increased economic benefits of expanding the design beyond its initial function, thus increasing the chance of generating sufficiently expanded economic benefits to exceed the economic costs. United States agencies such as the Army Corps of Engineers and the Bureau of Reclamation also used this requirement creatively in the practice of "river basin accounting," wherein economically questionable irrigation projects could be funded if the river basin also included profitable hydroelectric facilities (Robinson, 1979, pp. 25–27).

This unfettered expansion in U.S. dam construction, abetted by a favorable political-economic institutional setting, attenuated in the early 1970s with the advent of a burgeoning environmental movement and its associated environmental policies. With the passage of critical legislation such as the National Environmental Policy Act (NEPA) in 1970, the Clean Water Act (CWA) in 1972, and the Endangered Species Act (ESA) in 1973, environmental groups gained the necessary standing to voice opposition to dam construction, constrain ongoing projects, and abort planned projects. The greater scrutiny afforded by this legislation also coincided with the development of environmental economics as a discipline, which provided the requisite econometric tools and acumen to better evaluate and include the actual social and environmental costs of planned projects (Adler & Posner, 1999; Fisher & Krutilla, 1975; Pearce, 1998; Schulze, 1991). Many factors certainly coalesced to explain the reduction in dam construction in the United States (e.g., best sites for power generation had already been used, the increased opportunity costs of investing in large dams, etc.), but the incorporation of these environmental costs into traditional cost–benefit analysis in combination with key environmental legislation led to the general cessation of dam construction by the end of the 1990s, with only sixteen dams > 15 meters high being constructed since 2000 (Figure 6.1).

The global pacing of dam construction differs somewhat from U.S. trends. Global construction peaked about a decade after the U.S. construction, with a less pronounced declining tail. Large (> 15 meters high) dam construction continues at present, although the rate has diminished over the past twenty years (Figure 6.1). Unlike in the United States, global construction of new dams has seen a noticeable uptick over the past twenty years, especially in Europe and Asia, and, to a lesser degree, in non-U.S. North America (Table 6.1). Besides a general increase in the number of dams, these regions have also witnessed an increase in the number of especially large dams since the 1990s, principally in

Table 6.1 Dam construction by continent/region by decade

	< 1900 or no data	*1900–1909*	*1910–1919*	*1920–1929*	*1930–1939*	*1940–1949*	*1950–1959*	*1960–1969*	*1970–1979*	*1980–1989*	*1990–1999*	*2000–2009*	*2010–2019*	*Total*
Africa	140	15	15	37	48	48	143	271	383	566	438	117	54	2275
America C.	2		2		1			2	4		1	3	6	21
America N.	914	337	522	609	650	549	1358	2574	2012	1008	450	74	50	11107
America S.	269	11	80	80	96	68	192	226	206	223	132	210	84	1877
Asia	7,624	98	212	250	318	366	3105	4984	9686	4128	2445	1623	401	35240
Austral-Asia	41	17	25	38	38	24	66	109	115	149	105	23	2	752
Europe	604	135	135	224	296	259	767	1119	1075	966	703	449	262	6994
Total	9594	613	991	1238	1447	1314	5631	9285	13481	7040	4274	2499	859	58266

Source: Data from the World Register of Dams.

Table 6.2 Number of large dams (> 15 m) since 2000 for select countries

I.	*Asia*		
	2000–2009	2010–2015	Total
China	687	163	850
India	284	12	296
Iran	140	117	257
Japan	185	36	221
Korea (Rep. of)	114	2	116
II.	*Europe*		
Turkey	209	197	406
Spain	53	33	86
Portugal	38	4	42
Greece	58	8	66
France	21	3	24

Source: Data from World Register of Dams.

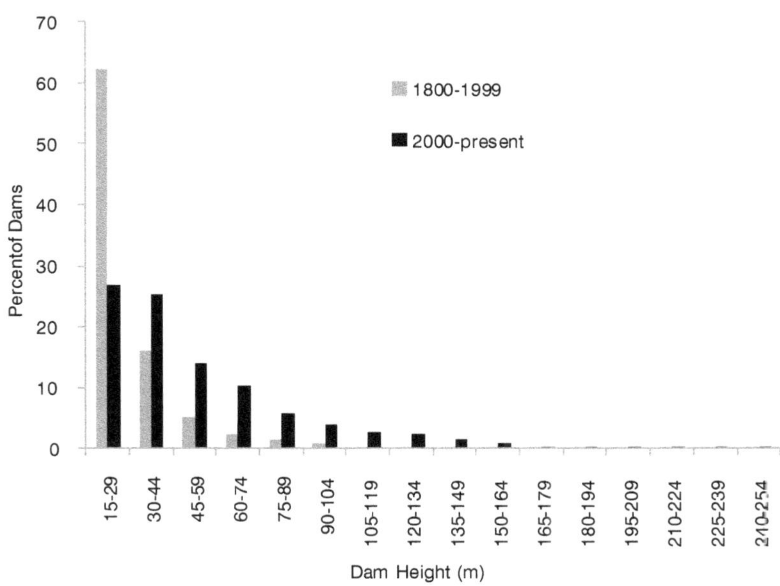

Figure 6.2 Heights of dams constructed globally prior to 2000 and since 2000.

Note the increased number of large dams (30–74 m) being constructed since 2000

Source: Data from World Register of Dams

Asia. Globally, prior to 2000, most dams (73 percent) were between 15 and 29 meters high. Since then, newly constructed dams range across a height spectrum (Figure 6.2), with a much greater number of very large dams boasting a height of between 30 and 104 meters (Figure 6.2).

A more spatially focused view indicates that the increase in recent large dam construction is limited to a handful of specific countries (Table 6.2). In Europe, Turkey, Greece, and, to a lesser degree, Portugal have been increasing their large dam construction since 2000. Between 1930 and 2000, Turkey only constructed 486 large dams. Since 2000, this pace has accelerated, with the erection of an additional 400 new large dams. Greece has also significantly accelerated the speed of dam construction; with only 66 dams constructed from 1920 to 2000, the country has now doubled the number in its dam portfolio, building 66 new dams since 2000.

On a regional scale, however, Asia by far accounts for the highest number of new large dams with China responsible for the majority of new dam construction, having added 850 dams since 2000. China has dominated and continues to dominate the regional census (23,840 total dams), with many of these dams built for either hydropower generation or irrigation. Moreover, the Chinese government has launched ambitious new river basin development programs within its territory and, as mentioned above, seeks to fund and erect countless dams in other parts of the world (Hensengerth, 2013). India (296), Iran (257), and Japan (221) have also accelerated new large dam construction since 2000 (Table 6.2).

Not only has Asia built the highest number of dams since 2000 of any region, it also boasts the greatest increase of dams with a significant ability to store water. Of the 2,242 dams with post-2000 reservoir volume data, most (2,222) are dams with less than 100,000 cubic meters of storage, with Asia accounting for the majority (1,542). Of the remaining 20 dams, 12 have reservoir storage capacities between 100,000 and 200,000 cubic meters, with Asia again having the most (7). Additionally, only Asia constructed dams with reservoir capacity exceeding 200,000 cubic meters (n=8). The ICOLD data set does not provide information on individual dam function, but it estimates that a large portion of new dam construction in China is dedicated to hydropower generation. According to ICOLD (2015), China generated 721 terawatt-hours (TWh) of hydroelectric power in 2010 (17 percent of its electricity consumption), with 2013 estimates of an additional 29 gigawatts of hydropower generation capacity.

Assessing the future construction of dams is particularly challenging, as data are difficult to assemble on an international level. This is especially the case in China, which will be the most significant locus of new dam construction both within its borders and beyond (see International Rivers, 2012). Because of its robust hydropower potential, many aid agencies and multinationals that perceive hydropower as a renewable and relatively inexpensive energy option are targeting Africa. According to a recent World Bank report, 93 percent of potential hydropower energy production in Africa is untapped (World Bank, 2009). Jim Yong Kim, president of the World Bank, also recently indicated that

Table 6.3 Current status of dams in Amazonia

Country	*In operation*	*Inventoried*	*Planned*	*Under construction*	*Total*
Bolivia	4		9	1	14
Brasil	74	56	90	30	250
Columbia		1			1
Ecuador	7	34	6	8	55
French Guyana	1		1		2
Guyana		1	30		31
Peru	8	36		3	47
Suriname	1			1	2
Venezuela	5				5
TOTAL	100	128	136	43	407

Source: Data from International Rivers, 2014. See also Hurwitz (2012).

the World Bank desires to expand hydropower production in the developing world and is targeting Africa as a major region as part of its new dam-building agenda: "In the case of places like Africa, we feel very strongly that they deserve energy and we need to move forward and find ways of providing that energy as cleanly as possible" (Kim, 2014). With its abundant water sources, the Amazon basin is also a locus of future large dam construction, and the recent controversy surrounding the Belo Monte Dam (see below) is representative of the types of political struggles bound to characterize dam building in this culturally and ecologically significant region. The advocacy organization International Rivers estimates that throughout the basin, over 400 dams are either in operation or are currently being constructed, inventoried, or planned (see Table 6.3).

Biophysical impacts and concerns over large dams

One of the central questions facing public officials, the energy industry, and the general public is whether or not the benefits of large dams are worth their substantial ecological costs. Considerable environmental assessments over the past three decades have revealed the significant effect of dams on ecological integrity (Junk, Bayley, & Sparks, 1989; Nilsson & Berggren, 2000; Nilsson et al., 2005). Because many of these projects can store considerable volumes of water—to be released gradually and at non-natural timings (Poff et al., 1997)—most of their ecological, hydrologic, and geomorphic impacts result from changes in the magnitude (high and low flows), timing, duration, and spatial extent of the modified flow regime. These modified hydrologic regimes significantly disrupt ecological function for both aquatic and terrestrial ecosystems, having tremendous impacts on predator–prey relationships, spawning cues, habitat requirements, reproduction cycles, and other life-cycle components (Ligon, Dietrich, & Trush, 1995; Magilligan & Nislow, 2001; Mahoney & Rood, 1998; Power, Dietrich, & Finlay, 1996; Wootton, Parker, & Power, 1996). River fragmentation and flow regulation by dams is now truly global in scope, with an estimated 48 percent of the planet's river volume moderately to severely affected by dams, a proportion that is expected to increase to 93 percent if all large dams currently under construction or planned are brought to fruition (Grill et al., 2015).

Dams, both those with storage capabilities as well as smaller, run-of-river facilities, fragment watersheds and disconnect the normative flux of energy, mass, and water—both longitudinally and laterally (Bunn & Arthington, 2002; Hughes, 1997; Kondolf et al., 2006; Magilligan, Nislow, & Graber, 2003; Vannote et al., 1980). Their mere presence acts to block migratory runs of diadromous species, especially salmon. Flow regulation and impoundment have contributed to pervasive reductions in U.S. salmon populations, especially in the Pacific Northwest, where numerous salmon runs are attaining endangered or threatened status under the Endangered Species Act. In the eastern U.S., Atlantic salmon have been extirpated from most drainages in New England, with only the Penobscot basin in Maine experiencing sustained runs.

The ecological effects of dams are not restricted to the stream channel, as the impacts of reduced flows can extend into the immediate riparian zone and across the broader floodplain (Bayley, 1995; Junk, Bayley, & Sparks, 1989; Molles et al., 1998; Nilsson & Berggren, 2000). Most riparian species can tolerate sustained flooding, and this adaptation serves to keep out more mesic–xeric upland species from invading. Reduction in the spatial extent of flooding by flow regulation have drastically affected floodplain floristic composition (Bayley, 1995; Junk, Bayley, & Sparks, 1989). This loss of plant diversity is especially acute in neo-tropical watersheds where broad lowland floodplains exist, often with sustained long-duration flows.

The targeting of these biomes for new dam construction represents a hotspot of concern about declining biodiversity. In particular, the ecologically diverse Amazon basin is targeted for a major initiative of large dam construction (Tundisi et al., 2014). Finer and Jenkins (2012) estimate that plans are underway for the construction of approximately 150 new dams over the next twenty years across five of the six major Andean tributaries of the Amazon (Table 6.3). Their assessment of the environmental impacts classifies 47 percent of the potential new dams as having "high impact" when considering the effects of flow regulation and other hydrologic shifts, with only 19 percent as generating "low impacts." When including the broader impacts of hydrologic fragmentation, this analysis of new dam construction indicates significant watershed dis-connectivity between protected headwaters in the Andes and the broader distal lowland Amazon basin. Moreover, when considering attendant construction activities, including new roads and transmission lines, Finer and Jenkins (2012) estimate that more than 80 percent of these proposed new dams would precipitate significant deforestation.

The politics of large dams

Given the significant biophysical impacts of large dams, ranging from local and regional projects to transnational basins transcending political boundaries, it's surprising that knowledge of these effects has not noticeably influenced decisions to build large dams. Likely, economic and political factors have simply overwhelmed any social and ecological concerns. Moreover, the decision to go ahead with a large dam project is further influenced by a host of political factors that seem to outweigh a careful scientific assessment of dam impacts. These dimensions include political and economic incentives for national governments to pursue large dams, resistance to projects from dam-affected communities and a variety of NGOs, and international tensions and hydropolitics that can arise when rivers shared by two or more sovereign states are targeted for a dam. In a much different vein, recent cases of dam removal to promote river restoration underscore the political character of dams throughout their lifetime. In keeping with the themes of this book, this section highlights the multiple ways in which large dams both result from and serve as catalysts for political processes on several spatial and temporal scales.

On one level, many countries' renewed emphasis on intensifying and hastening the transformation of river basins discussed previously is abetted by an international political economy that provides more and more fruitful ground for the implementation of large-scale water infrastructure. This renewed international push for dams is driven by several problematic assumptions: that the economic benefits of large dams outweigh their substantial social and environmental costs, that these benefits will be distributed equitably within and across diverse societal actors, and that hydropower is a form of renewable energy that will help combat the scourge of global warming. Recent research assessing the economic benefits of large dams finds that projects typically incur massive cost overruns and are frequently completed woefully behind schedule, making them very risky investments for developing countries (Ansar et al., 2014). Similarly, an economic assessment of projects in India concludes that the country's large dams have tended to exacerbate existing socioeconomic inequalities and, on average, increase poverty (Duflo & Pande, 2005).

However, many of the most important actors and institutions with the financial and political capacity to advance a global dam-building agenda—consisting largely of powerful states (e.g., China) and international financial institutions—perceive dams and hydropower as crucial elements of national economic development strategies. Governments as diverse as Iran, Turkey, Laos, Cambodia, Pakistan, Ethiopia, Brazil, and Peru—propped up by the pronouncements of international financial institutions, such as the World Bank, and the private sector's renewed interest in funding water infrastructure—share this view. Large dams are also promoted as sources of clean, renewable energy that can assist in global efforts to combat greenhouse gas emissions, a claim we take up in the following section. At the conclusion of the twentieth century, the World Commission on Dams (WCD) momentarily shifted the debate over large dams to focus on the question of whether or not dams actually provide the human development ideals they ostensibly promote (e.g., improved well-being, enhanced economic activities). The WCD also began to hold up the possibility of more just and sustainable non-dam alternatives. The report concludes that, although dams "have made an important and significant contribution to human development," they have frequently produced severe social and environmental impacts borne excessively "by people displaced, by communities downstream, by taxpayers and by the natural environment" (World Commission on Dams, 2000, p. 310). Today, only fifteen years later, the debate has shifted back. Instead of asking whether or not hydropower dams should be constructed, dam critics are now asking how to mitigate the negative impacts of projects seen as necessary and desirable.

Renewed momentum behind dam building and the associated political fallout is evident in Brazil's Belo Monte Dam, a case that underscores the trends highlighted in Tables 6.1 and 6.3. These tables reveal South America in general and the Amazon basin in particular as regional hotspots of large dam construction. First investigated in the 1970s, the Belo Monte hydroelectric project—located on the ecologically significant Xingu River, a prominent tributary of the Amazon—has been in the planning stages for over three decades and has generated heated political debate within Brazil and internationally (Klein, 2015). As is often the case with past and current struggles over dams, the project's proponents consist of a powerful coalition of government agencies, construction companies, and international industrial operatives who perceive hydroelectricity and dams as potent sources of development and symbols of national progress. Indeed, Brazilian politicians consider the Belo Monte Dam to be a "flagship project of national modernization and development" leading former President Luiz Inacio Lula da Silva ("Lula") to promise its completion "by fair means or foul" (Hall & Branford, 2012, p. 853). Opponents of the project comprise several indigenous groups, a national anti-dam social movement, and numerous international conservation and environmental organizations, who argue the dam's negative social and environmental impacts far outweigh its nominal benefits.

Construction of the Belo Monte Dam continues even though a cost–benefit analysis has revealed that the social and environmental costs will outstrip the dam's economic benefits (de Sousa Júnior & Reid, 2010). Indeed, anti-dam advocates have accused the Brazilian government of failing to consider (or taking lightly) the WCD's recommendations on improved decision-making processes around large dams, including greater transparency, increased participation by affected communities, and rigorous energy options assessment. The Belo Monte case underscores how politics at the national level are often linked to local opposition (in this instance by indigenous communities directly affected by the dam) and global-scale actors including both promoters (e.g., the international aluminum industry) and opponents (e.g., prominent conservation organizations).

On a second level, large dams such as the Belo Monte have, over the past several decades, fomented a wide array of social conflicts surrounding their construction in specific locales and raised questions about how to compensate dam-affected peoples for cultural disruption, displacement, and loss of livelihood. According to some estimates, dam projects have displaced 30–60 million people globally (McCully, 2001). Another 500 million have been detrimentally impacted by the downstream effects of damming, which include decreased agricultural production in former floodplains, decimation of fisheries, and seawater intrusion in river deltas (Richter et al., 2010). On the heels of these impacts, communities directly affected by dams have often mobilized effective and resilient campaigns to halt construction—as in the case of opposition to the Sardar Sarovar project on the Narmada River in India during the 1990s and 2000s (Khagram, 2004)—or receive just compensation for livelihoods devastated by completed projects—as in the case of the Pak Mun Dam in northeast Thailand during the same period (Sneddon & Fox, 2008).

As evidence of the capacity of large dam politics to transcend multiple scales, these instances of opposition to specific projects eventually coalesced into a global movement. The aforementioned WCD was the direct outgrowth of a growing refrain from environmental scientists, dam-affected communities, and transnational advocacy networks arguing that the cumulative negative impacts of large dams far overshadow their benefits. Over the course of the past decade, prominent anti-dam social movements in numerous countries—including the Philippines, Canada, Brazil, India, Malaysia, Thailand, Mozambique, and Ethiopia—have utilized research concerning the economic and cultural consequences of large dams to emphasize how large dams contribute to the physical displacement of entire communities, the destruction of river-based livelihoods, and the loss of culturally significant landscapes and places. Most recently, controversial dam projects in Laos, Ethiopia, Mozambique, Brazil, Chile, and other parts of the Global South have generated significant local opposition.

Historically, anti-dam campaigns have argued that many large dam projects are simply not worth the negative social and ecological impacts wrought by their construction. While not always successful in their efforts to halt construction, such campaigns have pointed out that many, if not all, benefits of large dams accrue to political and economic elites rather than contributing in any meaningful way to society's overall well-being. In the aforementioned Belo Monte case, at least some evidence shows that wider societal debates over the Brazilian project opened up heretofore non-existent democratic spaces for those most affected the dam—local fishing communities—to voice their message and be acknowledged by the national government (Klein, 2015).

Thirdly, the operation of existing hydropower dams and the proposed construction of hundreds of additional dams within transboundary river basins—river systems shared by two or more sovereign states—have led to increasing concerns about the hydropolitics of shared waters (Sneddon & Fox, 2006). In prominent river systems throughout the world, including the Nile, Mekong, Tigris–Euphrates, Amazon, and many others, upstream states have built (or are planning to build) large hydroelectric projects that threaten the integrity of rivers in downstream territories. The Mekong River basin offers a particularly instructive example of such hydropolitics. The $3.8 billion USD Xayaburi Dam in Laos, currently under construction, will be the first dam situated on the mainstream in the Lower Mekong basin. The dam will dramatically impact downstream countries and resource-dependent communities, which depend upon ecosystem services provided by a free-flowing river. The dam also paves the way for up to eleven more dams planned for the mainstream, which most scientific studies suggest would be ecologically (and likely socially) catastrophic for the lower basin and tens of millions of residents (see ICEM, 2010). The project is sparking basin-wide tensions between the downstream states of Cambodia and Vietnam, and the upstream governments of Thailand and Laos. All these riparian countries will benefit from the hydroelectric power generation, but in the process, state–state conflict over the proposed mainstream dams threatens to undermine the international agreement to govern the Mekong River basin in a "sustainable" fashion, encoded in a 1995 treaty signed by the four countries of the lower basin.

Yet the construction of Xayaburi and the planned development of the Don Sahong Dam—farther downstream on the Mekong River near the Lao-Cambodian border—has prompted widespread opposition among a growing coalition of community organizations, regional NGOs, and international advocacy organizations. This coalition, while originating in anti-dam advocacy organizations in Thailand (and to some extent Cambodia) in the 1990s and 2000s, has strived to recast the proposed damming of the Mekong River's main channel in terms of lost livelihoods for the basin's 70 million residents and potentially catastrophic ecological changes (Sneddon & Fox, 2015). In doing so, the coalition members are explicitly calling for a pan-Mekong identity on the part of people likely to be most affected by the proposed dams.

In stark contrast to the vision of the Mekong River basin adopted by governments and the global dam industry over the past five decades—wherein the Mekong, through its capacity to produce hydroelectricity and store water for irrigation, is seen exclusively as an engine of economic development—this bottom-up perspective puts forward a vision of the basin as a crucial source of livelihoods and as having intrinsic value. This "alternative Mekong" emphasizes, for example, the tremendously critical ecological services the river's flow provides in terms of fisheries production, which amounts to 2.6 million metric tons annually—an astounding 20 percent of the world's total inland fisheries output (Sneddon & Fox, 2012). As these observations make clear, the politics of damming the Mekong—an internationally and regionally critical freshwater system (Ziv et al., 2012)—are exceedingly complex, with social actors arguing their positions in ways that, at different times, privilege local, regional, national, and global scales.

In sum, motivations for constructing the world's large dams, past and present, represent significant public expenditures by governments hoping to promote economic development and modernization, and are always tinged with political calculation. Conversely, the implementation of large dams has contributed to political struggles at local, national, and transnational levels over the past four decades, a trend that is certain to continue in the future as more and more dams are promoted and brought online. These struggles are frequently driven by the political-economic interests of governments eager to take advantage of large dams' promised benefits (hydroelectricity and irrigation water), and are propped up by international financial institutions who perceive damming as an efficacious means of promoting economic development. However, in recent years, proponents of large dams have begun to voice novel arguments in support of the controversial claim that hydropower dams are a crucial tool in global initiatives to combat climate change.

Clean energy or green energy?

One of the most significant debates surrounding water, energy, and climate revolves around the question of whether or not large dam reservoirs—in both tropical regions and temperate zones—are a substantial source of methane and carbon emissions (Fletcher, 2011). Early studies revealed that reservoirs contribute to planetary greenhouse gas (GHG) fluxes, potentially at significant levels, because they eliminate a carbon sink (i.e., the flooded plant material, especially important in forested areas) and discharge greenhouse gasses into the atmosphere. These releases occur as the carbon in soils and submerged organic material breaks down and converts to carbon dioxide and methane (Fearnside & Pueyo, 2012; Louis et al., 2000). It is important to acknowledge that scientific understanding of reservoir emissions is swiftly progressing, incorporating previously unexamined biogeochemical dimensions of hydropower reservoirs (Goldenfum, 2012; Guérin et al., 2006). Yet this has not prevented advocates of hydropower development, particularly governments with prominent dam-building agendas such as China, India, and Brazil, from justifying these agendas with the claim that hydropower contributes to a global public good by lessening GHG emissions.

In the most recent thrust towards hydropower development, proponents from a variety of states as well as International Financial Institutions (IFIs) such as the World Bank are asserting that hydroelectric dams are a clean source of energy, more sustainable than the combustion of fossil fuels. Indeed, this has become a central claim. In fact, many countries' dam-building programs are moving forward under the assumption that hydropower dams are a crucial part of meeting carbon emissions goals under the Kyoto Protocol. In essence, the Protocol's Clean Development Mechanism (CDM), allows industrialized countries to offset their carbon emissions by investing in "clean" energy projects in developing countries. Article 7 describes the CDM as one of the "flexible mechanisms" that nations can use to offset emissions.

To simplify, industrialized countries can subsidize hydropower projects in less developed economies under the CDM because hydroelectricity production qualifies, in theory, as emissions reduction credits (certified emissions reductions, or CER). Developing countries can then use these credits to achieve their emission targets under the Kyoto Protocol and can sell them to industrialized countries as a way of financing hydropower development (Fearnside, 2015a; Mäkinen & Khan, 2010; Subbarao & Lloyd, 2011). Globally, hydropower dams are the second most significant type of project receiving financing through the

CDM (Mäkinen & Khan, 2010). In line with the trends discussed earlier in the chapter, the largest earners of CERs for hydropower development under the CDM as of 2013 are China (59.7 million CERs for 1374 dams), India (12.7 million CERs for 243 dams) and Brazil (12.6 million CERs for 111 dams). Globally, developing countries are claiming emissions reductions on 2049 dams, and over 50 percent of these qualify as large dams under the ICOLD criteria (Fearnside, 2015a).

The key assumption allowing hydropower to be identified as a global public good that mitigates the harmful impacts of climate change is that hydroelectricity is "clean" in terms of its GHG emissions. However, as noted above, the scientific evidence for this claim is beset by uncertainty, contradictory evidence, and lack of comprehensive data. In this case, the scientific knowledge of reservoir GHG emissions is being absorbed into the large dam debate. One way this occurs is via the CDM mechanism for assigning certain hydropower projects as eligible for financing. Any government can claim zero emissions for a hydropower dam if the ratio of installed capacity of the dam-to-reservoir area is greater than 10 watts/square meter (a ratio called the power density). In other words, dams able to produce large amounts of hydroelectricity while impounding relatively small reservoirs are good candidates for financing under the CDM.

Yet, as Fearnside (2015a) argues, the amount of GHG emissions from tropical reservoirs (carbon dioxide and methane) is highly uncertain, meaning the power density is a crude and imperfect measure for determining actual emissions from reservoirs. Moreover, the CDM Executive Board, the body that authorizes projects in developing countries to trade CERs, is dominated by representatives of countries that stand to benefit most from CDM projects. On a broader level, researchers with connections to large hydropower companies have typically dominated the scientific assessments used to investigate reservoir emissions (Fearnside, 2015b).

Recent debates about the contributions of large dams and their reservoirs to carbon emissions— and hence, global warming—unequivocally demonstrate that scientific examination of dam impacts has been far from neutral. When characterized as a contribution to the global public good (in the form of reduced GHG emissions), dams receive generous support and funding from international agreements such as the CDM under the Kyoto Protocol. Moreover, when hydropower dams are labeled as "clean" and "green" and a path towards low-carbon development, questions surrounding basin-scale environmental and social impacts become depoliticized and subordinated to governments' claims that these projects are contributing to a global public good, namely a cleaner atmosphere. While the trope of "clean hydropower" recognizes the array of biophysical disruptions apart from GHG emissions caused by damming rivers, dam proponents have successfully come to define those effects as collateral. The social and environmental impacts of large dams, this argument states, should be considered tangential to the larger goals of providing "renewable" electricity to underserved populations, many of which reside in the Global South.

Conclusion

Arguably, with the damming of the world's rivers, the twentieth century witnessed a "concrete revolution" (Sneddon, 2015). Ultimately the environmental and scientific revolutions in the 1960s and 1970s confronted this concrete revolution, which served to diminish the rate of large dam construction in many regions of the world. These trends have coalesced in the United States and elsewhere, and dam removal has become a viable and important tool of river rehabilitation. However, with new actors such as China emerging on the global scene, and with large funding agencies such as the World Bank refocusing their global strategies on energy development, there is little question that we are experiencing a global boom in damming. Primarily centered on hydropower development, this boom will incur long-lasting ecological and social impacts (Zarfl et al., 2015).

Looking toward the immediate future, large dams, which have already engendered political conflicts, will likely continue to generate controversy due to their wide-ranging impacts on freshwater systems and the people whose livelihoods depend on these systems. Moreover, despite recent claims by the World Bank and others that clean energy from hydropower generation will help alleviate global poverty and dampen the effects of climate change, some are cautioning that even in this new era, well-intentioned environmental mitigation efforts and broadly inclusive cost–benefit assessments, large dams may still be an economic and environmental burden—especially at local and regional scales. A recent evaluation of the true economic costs of large dams argues that in most countries where large hydropower dams are being proposed, the cost, when considered in absolute terms (including time to completion), greatly exceeds initial estimates, making such projects highly unprofitable and economically burdensome. Furthermore, when considering large-scale displacements of people as well as associated impacts on environment and livelihoods, the true costs of dams may not yield the proclaimed level of benefits. Thus, in terms of political contestation, the new era of large dam building may very much resemble the old era of dam building: it could be déjà vu all over again.

Acknowledgements

This research was funded in part by the National Science Foundation (BCS:1263519) and by the Rockefeller Center at Dartmouth College (Grant #061511). We thank Timi and Cat for their invitation to present our work, but more importantly, we thank them for their patience.

Works cited

Adler, M. D., & Posner, E. A. (1999.) Rethinking cost-benefit analysis. *Yale Law Journal*, 165–247.

Agnew, J. (2011). Waterpower: Politics and the geography of water provision. *Annals of the Association of American Geographers, 101*, 3, 463–476.

Ahlers, R., Budds, J., Joshi, D., Merme, V., & Zwarteveen, M. (2014). Framing hydropower as green energy: Assessing drivers, risks and tensions in the Eastern Himalayas. *Earth System Dynamics Discussion,* 5, 1521–1541.

American Rivers. (2014). Retrieved from http://www.americanrivers.org/blog/mapping-dam-removal-success/

Ansar, A., Flyvbjerg, B., Budzier, A., & Lunn, D. (2014). Should we build more large dams? The actual costs of hydropower megaproject development. *Energy Policy, 69*, 43–56. doi:10.1016/j.enpol.2013.10.069

Bayley, P. B. (1995). Understanding large river floodplain ecosystems. *Bioscience, 45*, 153–158.

Billington, D. P., & Jackson, D. C. (2006). *Big dams of the New Deal era: A confluence of engineering and politics*. Norman, OK: University of Oklahoma Press.

Bromley, C. (2001). Political and legal analysis of the rise and fall of western dams and reclamation projects, A. *University of Denver Water Law Review, 5*, 204.

Bunn, S. E., & Arthington, A. H. (2002). Basic principles and ecological consequences of altered flow regimes for aquatic biodiversity. *Environmental Management, 30*, 492–507. doi:10.1007/s00267-002-2737-0

De Sousa Júnior, W. C., & Reid, J. (2010). Uncertainties in Amazon hydropower development: Risk scenarios and environmental issues around the Belo Monte Dam. *Water Alternatives, 3*, 2, 249–268.

Dore, J., Xiaogang, Y., Li, K. Y., & Lebel, L. (2007). China's energy reforms and hydropower expansion in Yunnan. In L. Lebel, J. Dore, R. Daniel, & Y. S. Koma (Eds.), *Democratizing water governance in the Mekong region* (pp. 55–92). Chiang Mai, Thailand: Mekong Press.

Duflo, E., & Pande, R. (2005). Dams. National Bureau of Economic Research, Cambridge, MA.

Egan, T. (1998). "Heralding a new era, Babbitt chips away at harmful river dams." *New York Times*, July 15. http://www.nytimes.com/1998/07/15/us/heralding-a-new-era-babbitt-chips-away-at-harmful-river-dams.html.

Fearnside, P. M. (2015a). Emissions from tropical hydropower and the IPCC. *Environmental Science & Policy, 50*, 225–239.

Fearnside, P. M. (2015b). Tropical hydropower in the clean development mechanism: Brazil's Santo Antônio Dam as an example of the need for change. *Climatic Change*, 1–15.

Fearnside, P. M., & Pueyo, S. (2012). Greenhouse-gas emissions from tropical dams. *Nature Climate Change, 2*, 6, 382–384.

Finer, M., & Jenkins, C. N. (2012). Proliferation of hydroelectric dams in the Andean Amazon and implications for Andes-Amazon connectivity. *Plos One,* 7, 4, e35126. doi:10.1371/journal.pone.0035126

Fisher, A. C., & Krutilla, J. V. (1975). Resource conservation, environmental preservation, and the rate of discount. *The Quarterly Journal of Economics*, 358–370.

Fletcher, R. (2011). When environmental issues collide: Climate change and the shifting political ecology of hydroelectric power. *Peace and Conflict Monitor*, 5.

Goldenfum, J. A. (2012). Challenges and solutions for assessing the impact of freshwater reservoirs on natural GHG emissions. *Ecohydrology & Hydrobiology, 12*, 2, 115–122.

Gottlieb, R. (1988). *A life of its own: The politics and power of water.* San Diego, CA: Harcourt Brace Jovanovich.

Grill, G., Lehner, B., Lumsdon, A. E., MacDonald, G. K., Zarfl, C., & Liermann, R. C. (2015). An index-based framework for assessing patterns and trends in river fragmentation and flow regulation by global dams at multiple scales. *Environmental Research Letters, 10*, 1. 015001. doi:10.1088/1748-9326/10/1/015001

Grumbine, R. E., & Pandit, M. K. (2013). Threats from India's Himalaya dams. *Science, 339*, 6115, 36–37.

Guérin, F., Abril, G., Richard, S., Burban, B., Reynouard, C., Seyler, P., & Delmas, R. (2006). Methane and carbon dioxide emissions from tropical reservoirs: Significance of downstream rivers. *Geophysical Research Letters, 33*, 21 n.p.

Hall, A., & Branford, S. (2012). Development, dams and Dilma: The saga of Belo Monte. *Critical Sociology, 38*, 6, 851–862. doi:10.1177/0896920512440712

Hensengerth, O. (2013). Chinese hydropower companies and environmental norms in countries of the Global South: The involvement of Sinohydro in Ghana's Bui Dam. *Environment, Development and Sustainability, 15*, 2, 285–300.

Hughes, F. M. R. (1997). Floodplain biogeomorphology. *Progress in Physical Geography, 21*, 501–529.

International Centre for Environmental Management (ICEM). (2010). MRC strategic environmental assessment (SEA) of hydropower on the Mekong mainstream: Summary of the final report. Hanoi, Vietnam.

International Commission on Large Dams. (2015). ICOLD register of dams. Retrieved from http://www.icold-cigb.org/GB/World_register/world_register.asp.

International Rivers. (2012). The new great walls: A guide to China's overseas dam industry. Retrieved from http://www.internationalrivers.org/resources/the-new-great-walls-a-guide-to-china%E2%80%99s-overseas-dam-industry-3962.

Junk, W. J., Bayley, P. B., & Sparks, R. E. (1989). The flood pulse concept in river-floodplain systems. *Canadian Special Publication of Fisheries and Aquatic Sciences, 106*, 1, 110–127.

Kim, J. (2014). Transcript of press conference with World Bank Group President Jim Yong Kim in Manila. Retrieved from http://www.worldbank.org/en/news/speech/2014/07/15/transcript-of-press-conference-with-world-bank-group-president-jim-yong-kim-in-manila

Khagram, S. (2004). *Dams and development: Transnational struggles for water and power*. Ithaca, NY: Cornell University Press.

Klein, C. A. (1999). On dams and democracy. *Oregon Law Review, 78,* 641.

Klein, P. T. (2015). Engaging the Brazilian state: The Belo Monte Dam and the struggle for political voice. *Journal of Peasant Studies*, ahead-of-print, 1–20.

Kondolf, G. M., Boulton, A. J., O'Daniel, S., Poole, G. C., Rahel, F. J., Stanley, E. H.,… Cristoni, C. (2006). Process-based ecological river restoration: Visualizing three-dimensional connectivity and dynamic vectors to recover lost linkages. *Ecology and Society, 11*, 2, 5. Retrieved from http://www.csupomona.edu/~marshall/rs414/RS414_Readings/Kondolf_06_Riv_Restor_Connectivity.pdf.

Ligon, F. K., Dietrich, W. E., & Trush, W. J. (1995). Downstream ecological effects of dams. *Bioscience, 45*, 183–192.

Louis, V. L. S., Kelly, C. A., Duchemin, É., Rudd, J. W. M., & Rosenberg, D. M. (2000). Reservoir surfaces as sources of greenhouse gases to the atmosphere: A global estimate. *BioScience, 50*, 9, 766–775.

Magilligan, F. J., & Nislow, K. H. (2001). Long-term changes in regional hydrologic regime following impoundment in a humid-climate watershed. *Journal of the American Water Resources Association, 37*, 1551–1569.

Magilligan, F. J., Nislow, K. H., & Graber, B. E. (2003). Scale-independent assessment of discharge reduction and riparian disconnectivity following flow regulation by dams. *Geology, 31*, 569–572.

Mahoney, J. M., & Rood, S. B. (1998). Streamflow requirements for cottonwood seedling recruitment: An integrative model. *Wetlands, 18*, 634–645.

Mäkinen, K., & Khan, S. (2010). Policy considerations for greenhouse gas emissions from freshwater reservoirs. *Water Alternatives, 3*, 2, 91–105.

McCully, P. (2001). *Silenced rivers: The politics and ecology of large dams.* London: Zed Books.

Molle, F. (2009). River-basin planning and management: The social life of a concept. *Geoforum, 40*, 3, 484–494.

Molle, F., Mollinga, P. P., & Wester, P. (2009). Hydraulic bureaucracies and the hydraulic mission: Flows of water, flows of power. *Water Alternatives, 2*, 3, 328–349.

Molles, M. C., Crawford, C. S., Ellis, L. M., Valett, H. M., & Dahm, C. N. (1998). Managed flooding for riparian ecosystem restoration – Managed flooding reorganizes riparian forest ecosystems along the Middle Rio Grande in New Mexico. *Bioscience, 48*, 749–756.

Nelson, A. (2015). Balkan dam boom threatens Europe's last wild waterways. Retrieved from http://www.theguardian.com/environment/2015/feb/20/balkan-dam-boom-threatens-europes-last-wild-waterways.

Nilsson, C., & Berggren, K. (2000). Alterations of riparian ecosystems caused by river regulation. *BioScience, 50*, 9, 783–792.

Nilsson, C., Reidy, C. A., Dynesius, M., & Revenga, C. (2005). Fragmentation and flow regulation of the world's large river systems. *Science, 308*, 5720, 405–408.

O'Connor, J. E., Duda, J. J., & Grant, G. E. (2015). 1000 dams down and counting. *Science, 348*, 6234, 496–497. doi:10.1126/science.aaa9204.

Pearce, D. (1998). Cost-benefit analysis and environmental policy. *Oxford Review of Economic Policy, 14*, 4, 84–100.

Poff, N. L., Allan, J. D., Bain, M. B., Karr, J. R., Prestegaard, K. L., Richter, B. D.,… Stromberg, J. C. (1997). The natural flow regime. *Bioscience, 47*, 769–784.

Power, M. E., Dietrich, W. E., & Finlay, J. C. (1996). Dams and downstream aquatic biodiversity: Potential food web consequences of hydrologic and geomorphic change. *Environmental Management, 20*, 887–895.

Richter, B. D., Postel, S., Revenga, C., Scudder, T., Lehner, B., Churchill, A., & Chow, M. (2010). Lost in development's shadow: The downstream human consequences of dams. *Water Alternatives, 3*, 14–42. http://hdl.handle.net/10535/6634.

Robinson, M. C. (1979). *Water for the West: The Bureau of Reclamation, 1902–1977*. Public Works Historical Society, Kansas City, MO.

Schneider, H. (2013). World Bank turns to hydropower to square development with climate change. *The Washington Post*. Retrieved from http://www.washingtonpost.com/business/economy/world-bank-turns-to-hydropower-to-square-development-with-climate-change/2013/05/08/b9d60332-b1bd-11e2-9a98-4be1688d7d84_story.html.

Schulze, P. C. (1991). Incorporation of environmental damages into cost-benefit analyses: An introduction. *Bulletin of the Ecological Society of America, 72*, 1, 15–19. doi:10.2307/20167243

Smith, S. V., Renwick, W. H., Bartley, J. D., & Buddemeier, R. W. (2002). Distribution and significance of small, artificial water bodies across the United States landscape. *Science of the Total Environment, 299*, 1, 21–36. Retrieved from http://www.sciencedirect.com/science/article/pii/S004896970200222X.

Sneddon, C. (2015). *Concrete revolution: Large dams, Cold War geopolitics, and the US Bureau of Reclamation*. Chicago, IL: University of Chicago Press.

Sneddon, C., & Fox, C. (2006). Rethinking transboundary waters: A critical hydropolitics of the Mekong Basin. *Political Geography, 25*, 2, 181–202.

Sneddon, C., & Fox, C. (2008). Struggles over dams as struggles for justice: The World Commission on Dams (WCD) and anti-dam campaigns in Thailand and Mozambique. *Society and Natural Resources, 21*, 7, 625–640.

Sneddon, C., & Fox, C. (2011). The Cold War, the US Bureau of Reclamation, and the technopolitics of river basin development, 1950–1970. *Political Geography, 30*, 8, 450–460.

Sneddon, C., & Fox, C. (2012). Inland capture fisheries and large river systems: A political economy of Mekong fisheries. *Journal of Agrarian Change*, 12(2–3), 279–299.

Sneddon, C., & Fox, C. (2015). A genealogy of the basin: Scalar politics and identity in the Mekong River Basin. In E. S. Norman, C. Cook, & A. Cohen (Eds.), *Negotiating water governance: Why the politics of scale matter* (pp. 41–57). Farnham: Ashgate Publishing Ltd.

Subbarao, S., & Lloyd, B. (2011). Can the clean development mechanism (CDM) deliver? *Energy Policy, 39*, 3, 1600–1611.

Tundisi, J. G., Goldemberg, J., Matsumura-Tundisi, T., & Saraiva, A. C. F. (2014). How many more dams in the Amazon? *Energy Policy, 74*, 703–708. doi:10.1016/j.enpol.2014.07.013.

Vannote, R. L., Minshall, G. W., Cummins, K. W., Sedell, J. R., & Cushing, C. E. (1980). River continuum concept. *Canadian Journal of Fisheries and Aquatic Sciences, 37*, 130–137.

White, G. F. (1957). A perspective of river basin development. *Law and Contemporary Problems, 22*, 157–187.

Wootton, J. T., Parker, M. S., & Power, M. E. (1996). Effects of disturbance on river food webs. *Science, 273*, 1558–1561.

World Bank. (2009). Directions in hydropower, Washington, DC.

World Commission on Dams. (2000). *Dams and development: A new framework for decision-making: The report of the World Commission on Dams*. Earthscan, London.

Worster, D. (1985). *Rivers of empire: Water, aridity, and the growth of the American West*. Oxford: Oxford University Press.

Zarfl, C., Lumsdon, A. E., Berlekamp, J., Tydecks, L., and Tockner, K. (2015). A global boom in hydropower dam construction. *Aquatic Sciences,* 77, 1, 161–170.

Ziv, G., Baran, E., Nam, S., Rodríguez-Iturbe, I., & Levin, S. A. (2012). Trading-off fish biodiversity, food security, and hydropower in the Mekong River Basin. *Proceedings of the National Academy of Sciences, 109,* 15, 5609–5614.

7 Oil and water don't mix

Impacts of oil and gas development on Sakhalin's water resources

Jessica K. Graybill

Introduction

Environmental degradation in Russia is well documented by scholars who relate the ongoing water, air, and ground pollution to the former Soviet Union's widespread, fast-paced industrial development and its accompanying urbanization (e.g., Glavin, 2006; Pryde, 1991; Weiner, 2000; Wood & French, 1989), which continues to a certain extent in the Russian Federation. Not by coincidence, most of the Soviet era's development schemes—and thus environmental catastrophes—occurred in less populated and geographically peripheral regions such as in Ukraine (Chernobyl nuclear power plant explosion), Uzbekistan and Kazakhstan (Aral Sea depletion and salinization of agricultural lands), Chelyabinsk (home to a factory for plutonium production for nuclear power), Norilsk (nickel mine and smelter), and countless places in Siberia, the Arctic, and the Far East, where extraction of heavy metals and hydrocarbons continues. To build the communist dream and advance the region economically, politically, and socially, plans to industrialize and urbanize the entire Soviet Eurasian landmass meant exploiting rich natural resources in the peripheral regions.

What binds the industrial and urban developments above is water. In many ways, Russia's abundant freshwater resources enabled the pace and scale of its industrial development. Without access to freshwater resources, Soviet and post-Soviet Russia's industrial development arguably would have occurred more slowly and in less peripheral regions. Where water resources have been less plentiful, water-based megaprojects supplied water to industrial and urban sites through canals (e.g., to the Ferghana Valley from the Aral Sea; accomplished at great economic and environmental cost) or via the re-direction of major rivers (e.g., the Ob; never accomplished). Vast quantities of fresh water went to cooling nuclear production facilities, irrigating fields, sluicing mineral deposits, and hydrating and cleansing industrial workers. Commercial-scale fishing has replaced subsistence fishing in all fresh waters. Perhaps most importantly to economic development, Russia's waterways have been instrumental in transporting raw and industrial goods out of peripheral production regions to hubs for further transportation or consumption. In this way, fresh water constitutes a resource of a different kind: water acts as a conduit for the removal of goods to markets on, through, and away from the Eurasian landmass.

While freshwater pollution is widespread and commonly linked with industrial development (Glavin, 2006; Pryde, 1991), local citizens' concern for harm done to freshwater resources historically and today has been mild at best. Some scholars attribute a general lack of conflict over environmental degradation in Russia to a variety of factors, including a history of repression of free speech against the government's actions (Zimmermann, 2014), general disinterest in environmental topics except by special groups and clubs (Henry, 2010), a lack of knowledge of the extent of environmental degradation (Graybill, 2007), and weak civil society development (Henderson, 2011). However, other scholars have found that interest in environmental issues captures the hearts and actions of residents in many individual localities in the post-Soviet era. In a survey of social movements in multiple communities across Russia, political scientist Laura Henry (2010) found that local involvement in various kinds of environmental activities, from educational clubs to activism, speaks to a growing concern about environmental and social wellbeing in Russia during the 1990s and early 2000s. The ongoing, well organized, and politically backed movement protesting the development of a transportation corridor through the Khimki Forest near Moscow also speaks to the ability of Russian citizens to mobilize around environmental topics when social and political conditions permit (Boentje & Blinnikov, 2007; Evans, 2012).

Sakhalin Island in the Russian Far East has seen open and expressive contestation of harm to freshwater resources in the post-Soviet era (Figure 7.1). While onshore oil extraction has occurred on Sakhalin Island for about a century, only in the mid-1990s did rapid exploration and plans for the development of offshore deposits begin to occur. The post-1991 introduction of market forces and joint venture capitalism in Russia coincided with the advent of new horizontal drilling technologies for offshore hydrocarbon extraction, making Sakhalin Island's reserves both available and tempting to transnational hydrocarbon corporations at an exciting time in their historical development of offshore, cold water extraction technologies (Gustafson, 2013; Thornton & Ziegler, 2002). Onshore infrastructure plans and actualities (e.g., roads, camps, export terminals) wreaked havoc with the local environment, and critique of hydrocarbon development plans for Sakhalin began to surface in the mid-1990s, coinciding with the dawn of more open outcries across Russia against environmental and social ills in the country, as noted by Henry.

However, contestation of development due to environmental and social concerns seems to have been limited to the period when transnational hydrocarbon corporations performed the majority of development, from the mid-1990s until 2006. After 2006 and the (re)turn of the majority of offshore development to Russian state corporations, the short-lived yet highly critical and global campaign against freshwater degradation on Sakhalin Island lost its strength, leaving a question mark about the possibilities for future environmental movements in this locale (Bradshaw, 2006; Graybill, 2013).

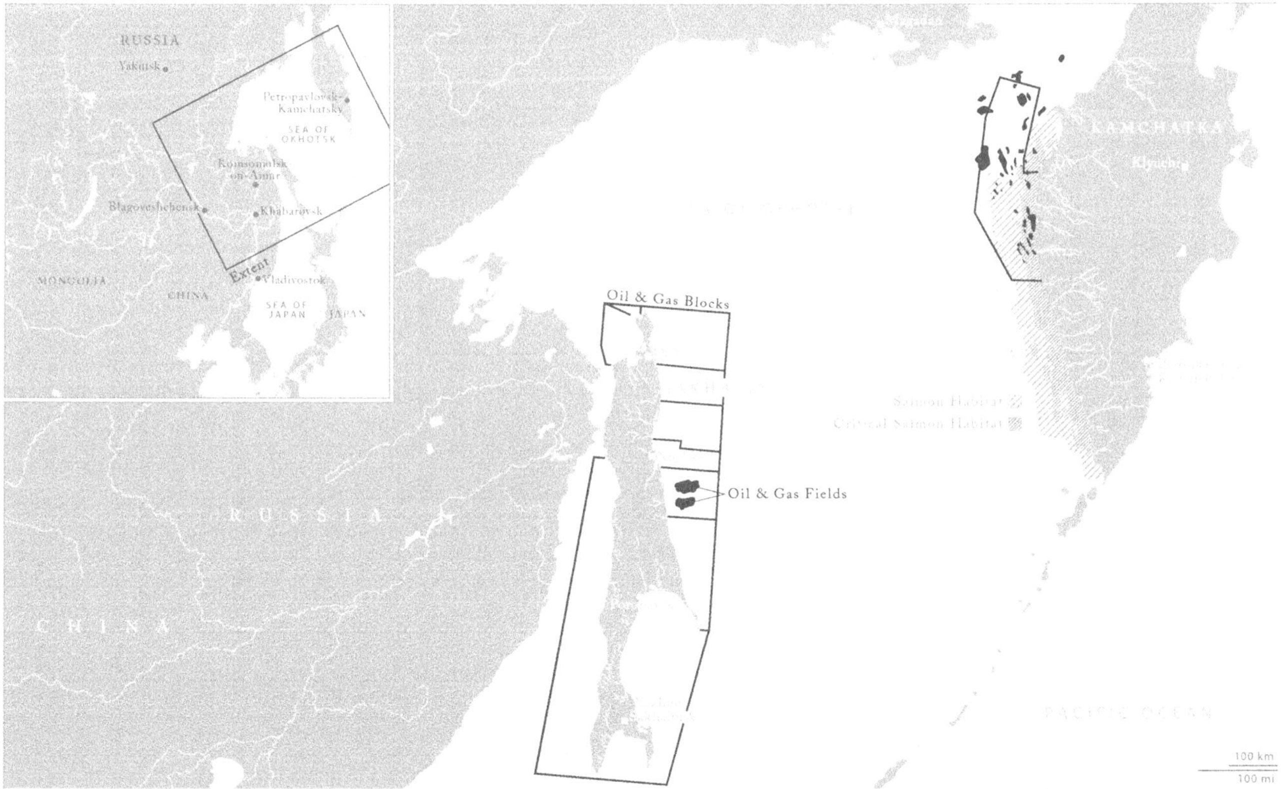

Figure 7.1 In the northern Sea of Okhotsk the coastal regions of Sakhalin Island and western Kamchatka are critical meeting points for marine and riverine waters and for different resources extracted from the region, namely salmon and offshore hydrocarbons (oil and gas).

Source: Cartography by Joe Semanchik and Gregory Woolston using data from Rand et al., (2012), Zhbanbova, P. (2013), and Natural Earth.

Sakhalin's polluted peripheral waters

Across the Russian Far East, low human population densities stretch across seemingly endless tundra and taiga ecosystems from the eastern extent of Siberia to the Pacific Ocean and from the southern Chinese border to the northern Arctic Ocean. Archipelago-like population centers created during the Soviet era largely cluster in southern Russia along the Chinese border, the Trans-Siberian Railroad, and the coast of Pacific Russia (Stephan, 1994). Many places, especially more northern ones, remain inaccessible by railroad or by permanent road and are experiencing rapid population outmigration, in some cases devastating urban and rural settlements culturally and socioeconomically (Heleniak, 1999). For example, the Trans-Siberian Railroad hugs the southern border and ends in Primorye, leaving a vast swathe of northern Russia approachable in only a few ways and only for a few months of the year. Winter affords the greatest mobility here, where dogsleds or sledges serve for transport over frozen rivers and, more recently, where ATVs (all-terrain vehicles) provide transport over snow and ice roads. Only in the past thirty years has air travel been viable for human and resource mobility through this region, and then only for a limited number of people and at a price.

The Russian Far East has been perceived as a space of multiple abundant resources by indigenous groups, early explorers, and early Slavic settlers (such as the Old Believers), Tsarist colonizers, and Soviet and post-Soviet newcomers (Stephan, 1994). Historically, waterways, both riverine and marine, transported important resources out of this region. Many resources, including furs, precious minerals, metals, and fish, flowed out and away by ocean vessel via the region's northward flowing rivers into the Arctic Ocean and around to Europe and Asia.

Beginning with the Mongol and Chinese empires (Stephan, 1971; Vysokov, 1996), Sakhalin Island, located in the Sea of Okhotsk in far eastern Pacific Russia, has long been perceived as a resource periphery, and it has served as a colonial outpost of various national entities since about 900 AD. In more recent history, the Japanese claimed Sakhalin as an extension of the Japanese Archipelago, but Russians took possession of the island in 1852, and it officially became subject to Russian rule in 1875 (Stephan, 1971). The writer Anton Chekhov, who visited Sakhalin in 1895, immortalized it in *A Journey to Sakhalin* as one of the harshest penal colonies to dot Siberia and the Far East in the Russian Empire's late period. In unrelenting detail, Chekhov documents the depressing lives and sordid living conditions of hard labor prisoners performing hard labor, largely in Sakhalin's coal mines.

Major resources extracted from Sakhalin historically and at present have included minerals, coal, oil, gas, timber, and salmon (Newell, 2004; Wood & French, 1989). Perhaps Sakhalin's most important and longstanding resource has been salmon. Multiple regions in the Russian Far East are prime habitat for salmon spawning, and scholars (Augerot, 2005; Rand et al., 2012) and activists (e.g., from Pacific Environment, Wild Salmon Center) claim the Russian Far East as the last great frontier for wild salmon today. On Sakhalin, the combination of shallow nearshore marine basins, cold-water temperatures, and topographically easy inland access to thousands of freshwater streams enables the anadromous Pacific salmon to thrive.

Regarding oil and gas resources, the geologic characteristics of the rocks and sediments across the island have created rich conditions for onshore and offshore deposits of precious metals (e.g., platinum) and offshore oil and gas deposits (Zonenshain, Kuzmin, & Natapov, 1990). Poorly executed and managed exploitation of onshore oil and gas in northern Sakhalin, near the city of Okha, has led to massive environmental degradation due to oil seepage from hundreds of derricks surrounding the outskirts of this city. Personal observation in 2005 indicated unmitigated pollution of the environment, evidenced by puddles of oil and other extraction compounds around Soviet-era derricks that slowly and continuously leaked into the ground and small waterways that in turn carried the toxic oil mixtures out to the Sea of Okhotsk.

Increasingly, the objects in Sakhalin's freshwater resources constitute a source of concern for people and the environment in this resource periphery. The presence of industrial waste from resource extraction is well known in Sakhalin's waterways, especially in the multiple bays surrounding the island. Of particular concern are heightened concentrations of petroleum hydrocarbons, phenols, and heavy metals (iron, lead, and cadmium) in the water and bottom sediments (Leonov & Pishchal'nik, 2005; Polteva, Latkovskaya, & Leonov, 2009). Industrial waste from mineral and hydrocarbon extraction processes remains from the Soviet period of Sakhalin Island's industrialization but has been increasing since 1991 as a result of exploration and extraction (Leonov, 1999).

Fresh water, salmon, and subsistence survival

Multiple indigenous groups populated Sakhalin for thousands of years prior to the arrival of the first Asian and European explorers in this region (Slezkine, 1996). Russian ethnographer Lev Sternberg and self-trained cultural anthropologist Bronisław Piłsudski, both of whom lived on Sakhalin at the turn of the twentieth century while researching and writing about native peoples and their cultures, provided the first reports of the Ainu, Nivkh, and Evenk's rich indigenous cultures. Tradition holds that the Ainu, Nivkh (Gilyak), Nanai, and Uilta were the first true inhabitants of Sakhalin, relying on a fish-centered subsistence culture in marine and riverine environments. The Evenk arrived on Sakhalin later (Slezkine, 1996; Vysokov, 1996), introducing reindeer herding culture from the Russian mainland to the island.

Today, Sakhalin claims an indigenous population of approximately 4,000, of which Nivkh comprise 2,682 residents, Uilta 362, Evenk 326, and Nanai 164. The Ainu culture on Sakhalin was decimated after the Russo-Japanese War of 1905 and again after Soviet takeover of the entire island in 1946; the remaining Ainu inhabit Hokkaido, Japan (Slezkine, 1996). The indigenous population today constitutes approximately 0.7 percent of the island's population, and indigenous settlements are largely relegated to smaller and less industrialized regions within Sakhalin and the outskirts of urban settlements dotted across the island. Originally, these indigenous groups were semi-nomadic and nomadic (Evenk only), moving with the fish from downstream to upstream camps. Their traditional subsistence lifestyles ended with Soviet collectivization in the 1930s and 1940s. During this time, the Soviet state relocated indigenous people to compact urban-like settlements and, as elsewhere in Soviet Russia, indigenous children attended boarding schools (*internaat*; Bartels & Bartels, 1995) in the state's colonial attempt to weaken traditional cultures and strengthen allegiance to the Russian language and Soviet culture (Vakhtin, 1992).

During the Soviet period, indigenous peoples enjoyed access to the same educational opportunities offered to all Soviet citizens. However, because of the state's traumatic disruption of traditional culture, language, and kinship networks, and because of the placement of indigenous collectivized settlements in out-of-the-way locations (i.e., away from urban centers and on the edges of Soviet industrial efforts), many indigenous peoples did not, or could not, take advantage of educational and social betterment projects. Many completed only the minimum schooling required and returned to their now-collectivized and displaced, yet still "traditional," homelands to work among fellow indigenous peoples and Soviet newcomers in small industries such as collective farms or fishing and canning collectives (Grant, 1995). Some completed higher educational degrees in St. Petersburg in ethnic studies fields, thereby allowing for the preservation of traditional crafts (e.g., embroidery, skin curing) and arts (e.g., dance, music) to the present day (personal communication, Nogliki Museum Director).

Largely, the decision of many indigenous residents to remain near their traditional homelands ensured the continuation of their connection to Sakhalin's terrestrial and water-based resources. Indigenous peoples continued traditional hunting, fishing, and gathering activities throughout the Soviet period, a practice that supplemented their diets during lean years of Soviet industrial development. During the post-Soviet period, increased traditional activities not only supplement diets but also the economic circumstances of those left far behind after the disappearance of Soviet-era subsidies in this peripheral region of Russia. One of the most important fishing activities in Sakhalin's fresh waters today is the legal and illegal catch of salmon, largely for the purposes of removing roe for market sale (Graybill, 2008; Newell, 2004). Thus, freshwater environments are of crucial importance to local and indigenous peoples, many of whom depend on the health of rivers—and the fish that run in them—in order to stay economically afloat.

Interviews with residents of Sakhalin conducted between 2003 and 2006 revealed that both local and indigenous peoples held strong emotions about perceived changes to water quality and to the quality of the fish in the water. The most prominent emotions displayed were love for and awe of place, despair at the loss of ecological habitat (especially of salmon), fear of hydrocarbon infrastructure, and fear of re-appropriation of cultural and ecological space (Graybill, 2013). Many people perceived the quality of water—and thus the quality of salmon—to be changing, noting, "The fish already taste like oil" (Graybill, 2008).

Conflict rising

The extractive relationship of Sakhalin Oblast (an oblast is an administrative unit, similar to states) to multiple empires—Mongol, Chinese, Japanese, Russian, Soviet, and again Russian—indicates its historical treatment as a periphery from which distant power centers extracted resources. Treatment of indigenous peoples and, later, Soviet newcomers as socioeconomically and culturally peripheral has created a local culture that perceives exploitation of Sakhalin's natural resources as the only viable option for economic survival. According to one local resident, an extractive relationship with Sakhalin's environment is "only natural" in this relatively pristine place (Graybill, 2009).

Locating Sakhalin's place in geographic history and discussing its cultural makeup is important for understanding how contemporary indigenous and local people perceive and construct environmental problems, especially those associated with their continued and vital subsistence resource use that stems from ruralization, a common modus operandi since the collapse of the Soviet Union (Ioffe & Nefedova, 1998). Because people depend on local terrestrial (e.g., fern, mushrooms, herbs, wild roots) and marine (e.g., salmon, seaweed, fish, crabs, scallops) resources for everyday sustenance (Graybill, 2009; Wilson, 2003), attachment to these resources is strong among indigenous and local communities who struggle with ongoing socioeconomic and political transformation.

Local and indigenous people in Sakhalin Oblast are concerned about the environmental integrity of their ecological homelands because their ability to survive and thrive in the post-Soviet era has become increasingly dependent on subsistence resources. Additionally, local and indigenous communities fear that the future of their cultural identities, including, for instance, their ability to (re) build past traditions such as traditional salmon harvesting, may be under threat by socioeconomic globalization in the twenty-first century (Grant, 1995). Thus far, major beneficiaries from regional industrial development include the federal and regional governments, individual regional authorities (e.g., governors), and transnational corporations. Long-term, tangible benefits from upstream and midstream hydrocarbon development have not reached local and indigenous communities (Bradshaw, 2006; Meier, 2000; Rutledge, 2004; Wilson, 2003). Even when socioeconomic impact assessments related to resource development have included these communities, as was the case for development of Sakhalin's projects (Bradshaw, 2006), residents perceive their inclusion to have been minimal (Graybill, 2008; Wilson, 2003) and their desires and concerns only superficially addressed (Graybill, 2009).

As ruralization was beginning to increase indigenous and local dependence on Sakhalin's resources, the collapse of the Soviet Union in 1991 opened the doors for transnational oil companies in Russia's sub-Arctic and Arctic marginal ocean basins (Gustafson, 2013). Generally seen as one of the last frontiers for large-scale hydrocarbon development, this vast, underdeveloped region has been of great interest nationally and internationally as a hydrocarbon resource development area. Because of the extreme offshore conditions for extraction (e.g., under sea ice, using horizontal drilling technology), engineering ingenuity in the use of offshore underwater extraction technology is necessary for extraction. The advent of joint venture operations between transnational corporations (TNCs) and Russian national corporations in the mid-1990s is the main reason that mineral, oil, and gas resources flow in, through, and out of this peripheral region today, as foreign investments and foreign technology fuel resource extraction in the offshore region.

Among the first regions to move forward with large-scale exploration and extraction by TNCs was Sakhalin (Rutledge, 2004). Due to Sakhalin's location in the largely undeveloped and environmentally fragile sub-Arctic region, development of Sakhalin's offshore hydrocarbon reserves is environmentally risky. Risks include damage to terrestrial, riverine, and nearshore environments from hydrocarbon infrastructure emplacement (e.g., building of roads, substations, and a liquefied natural gas plant), all for the expectations of long-term (more than fifty years) hydrocarbon extraction from multiple offshore reserves in the Sea of Okhotsk (personal communication from Yuzhno-Sakhalinsk Urban Planning Department).

Today, offshore oil and gas extraction fuel Sakhalin's resource-based economy. Two major oil and gas fields are in production (Sakhalin-1 and Sakhalin-2), and exploration of other offshore fields continues. Sakhalin's offshore projects remain unique in the Russian oil and gas landscape because multinational joint-venture companies operating under production-sharing agreements (PSAs) between the Russian government and foreign partners in the Sakhalin-1 and Sakhalin-2 projects are providing the capital and technological investments necessary for extraction (Rutledge, 2004). Consortiums of Russian national companies and TNCs govern Sakhalin-1 (Exxon Mobil, Sakhalin Oil & Gas, OGNC Videsh, and Rosneft) and Sakhalin-2 (current composition: Gazprom, Shell, Mitsui, Diamond).

Sakhalin-2 is the first PSA ever signed in Russia, and its development has been fraught with environmental politics since the signing in 1994. In this agreement, the Russian government profits from hydrocarbon development only after all investors recoup most of the hydrocarbon production costs. Because Sakhalin Energy's projected costs more than doubled to at least US$22 billion (Rutledge, 2004) since the first estimates of the cost of infrastructure development for extraction in the mid-1990s, the Russian government came to understand by the mid-2000s that it would have to wait quite some time to begin profiting from this joint venture. Sakhalin-1's PSA is less problematic for Russia than Sakhalin-2's because of its particular wording, which yielded profits to the federal government at an earlier stage (Bradshaw, 2006; Rutledge, 2004).

Sakhalin-2 was Russia's only large-scale energy project operating without a Russian partner until 2006. Initially, Shell Corporation owned a majority stake in Sakhalin-2, and the corporation sought funding for hydrocarbon development from large-scale lending sources sometimes involved with development (e.g., European Bank for Reconstruction and Development). Because of Shell's involvement and that of large-scale funding agencies,

Shell investigated the social, economic, and environmental impacts of hydrocarbon development in this underdeveloped resource periphery, as did nongovernmental organizations (NGOs) concerned with this development's environmental, social, and cultural impact.

NGOs indeed voiced the loudest of the outcries against development's environmental and social impacts (Graybill, 2009). Well funded international NGOs, namely Pacific Environment, Wild Salmon Center, and Gender Action, worked both to create and to strengthen the work of local NGOs (such as Sakhalin Environmental Watch) and indigenous groups on Sakhalin Island (local units of the Russian Association of Indigenous Peoples of the North, RAIPON), empowering them to protest further oil and gas development without proper documentation of environmental and sociocultural impacts and without proper compensation for loss of traditional livelihoods due to landscape (terrestrial and freshwater) degradation. Women's health and wellbeing during the phase of infrastructure development fell under criticism as well (Bacheva, Kochladze, & Dennis 2006). These groups organized well and widely broadcasted their messages, offering stinging critiques of the pace and mode of development pursued by TNCs on Sakhalin Island proper and in the offshore drilling environments.

Researchers and activists critiqued environmental and social–cultural impacts of hydrocarbon development, aiming to reduce doubt about the "ills" of hydrocarbon extraction in this region during the post-Soviet era by TNCs and foment a movement to halt current practices of TNC development on Sakhalin. These researchers and activists documented the degradation of multiple habitats for multiple species of freshwater salmon (catalogued extensively on a local NGO's website), terrestrial tundra for caribou grazing (Wilson, 2003), and the nearshore spawning grounds of the Korean Grey Whale (documented in multiple places on Pacific Environment's website). They also documented the acres of forests along the eastern shore of Sakhalin near Nogliki, razed to build pipelines to carry hydrocarbons to Russia's first liquefied natural gas (LNG) export terminal in Prigorodnoye in southern Sakhalin. Additionally, researchers and activists documented the economic importance of terrestrial, riverine, and nearshore subsistence resource use by indigenous peoples (Graybill, 2013; Wilson, 2003) and the impact on local women of migrant workers helping to construct the hydrocarbon infrastructure (Bacheva et al., 2006). Protests against oil and gas development ensued, both locally and in major world cities (Moscow, London, and New York; Bradshaw, 2006).

The researchers and activists' work to raise concern about the mode of development on Sakhalin Island in the post-Soviet era indicates that TNCs conducted most of the economic development in this region and that development led to the TNCs' international and national-level economic benefit at the expense of local environments and local and indigenous peoples who, according to the PSAs drawn up for both Sakhalin-1 and Sakhalin-2, would receive only trickle-down benefits—and only after recouping the development's initial costs (Bradshaw, 2006; Graybill, 2009; Rutledge, 2004; Wilson, 2003).

Table 7.1 Human responses to the impacts of hydrocarbon development on the life stages of the region's critical resource, salmon

Extractive industries	*Activity*	*Salmon life cycles (lifespaces)*	*Timing of extraction process or of salmon growth*	*Geographic location: possible environmental effects*	*Human response to changing environments (social, political, economic)*
Upstream	Heavy metal/ mineral mining; onshore oil extraction	Freshwater (upstream)	Early	Upstream: degradation (water blasting, seepage, tailings)	Illegal salmon overharvest Protesting extractive industry
Midstream	Resource transport, storage	Marine (downstream)	Mid	Upstream: spills and pollution from road transportation across rivers, open-pit storage of byproducts	Policy creation and negotiation: catch amounts, pollution control measures
Downstream	Refining, marketing, distribution	Freshwater (upstream)	Late	Upstream: onshore processing (heavy water use) Downstream: spillage from/sinking of tankers in nearshore marine environment	Illegal salmon harvest

Of all the environmental and sociocultural impacts, environmental NGOs and biologists on Sakhalin voiced the most concern about the island's freshwater and nearshore marine environments (e.g., Leonov & Pishchal'nik, 2006; Polteva et al., 2009). Of greatest concern to indigenous and local peoples is the health of salmon habitat and salmon populations: without this vital resource, subsistence and small-scale economic survival fall under threat (Graybill, 2013). Table 7.1 indicates hydrocarbon development's impact on Sakhalin salmon populations, as documented by researchers and activists between the 1990s and the mid-2000s.

Interestingly, while Sakhalin's hydrocarbon development continued to rely on funding from TNCs and international lending, the door opened for concern about the impacts of hydrocarbon development on multiple environments and on people. Opportunity presented itself to bring attention to the island and to incite local–global contestation of the development, especially development conducted by Sakhalin-2. What remained unforeseen by those most determined to make heavy cases against Sakhalin's development was the Russian state's growing power to assume new control over the future of oil and gas development in the region. Indeed, in late 2006 the federal and regional governments criticized Sakhalin-2 for environmental degradation and violation of Russian environmental laws (Bradshaw, 2006). Citing environmental degradation by the TNCs in control of Sakhalin-2, the federal government issued a federal mandate, halting Sakhalin-2's operations in September 2006. Negotiations to resume work ensued, continuing without success until December 2006, when Shell's majority interest (50 percent plus one share) was ceded to Gazprom, Russia's state-owned gas company.

In this way, political maneuvering by the Russian government wrested control of hydrocarbon development away from TNCs and secured national participation in this lucrative regional hydrocarbon development. This trend is not unique to Sakhalin: the Russian state also renegotiated energy development deals made in the 1990s with other foreign investors in other Russian resource peripheries (e.g., BP-TNK's investment in the Kovykta Field near Irkutsk; Boykevich, 2006). In addition to securing Russia's national stake in projects with foreign investment—ostensibly in moves to shore up national security interests—President Putin continues to criticize, control, or remove from power regional leaders whom he perceives as becoming too powerful and influential or with whom he has conflict. For example, in March 2015, the governor of Sakhalin Oblast, Aleksander Khoroshavin, was arrested on charges of taking bribes associated with development of a regional heating plant. While official news sources convey that his arrest relates to new attention to corruption crackdowns in Russia, political pundits ask whether the arrest might be more appropriately understood as a message that "there are no untouchables" (Sinelschikova, 2015) in Russia, including powerful regional authorities who oversee governance of regions containing natural resources deemed important to national security.

Post-2006: conflict without teeth

Until 2006, both local and global environmental and cultural activists widely and continually publicized the struggle over resources, both hydrocarbon and subsistence (Bradshaw, 2005). Since 2006 and the transfer of majority holding in Sakhalin-2 to Gazprom, international financing institutions have refused to finance this project (i.e., the European Bank for Reconstruction and Development; Williams, 2007). Previously, environmental impact assessments written for international lenders ensured that either the Russian government or transnational operations would address environmental concerns. Previously, many indigenous activist groups also took encouragement from their newly found—and newly heard—voices, requesting "cultural impact assessments" in protests against extraction projects (Graybill, 2009). This checks and balances approach, alongside local–global environmental and indigenous activism, helped guarantee that hydrocarbon development followed international standards. Since Gazprom's majority stake takeover, discussions and knowledge of environmental conditions outside the region have disappeared; some speculate that greening of the development process is unlikely to increase with the Russian major as the main operator (Bradshaw, 2006). It remains to be seen whether further greening—or even browning of this project—will occur and whether environmental and cultural activists will regain the power they held in the early 2000s.

Renationalization of hydrocarbon development on Sakhalin leads to questions about how multiple stakeholders in the future may deal with environmental degradation in this and other Russian regions, especially regions with vital freshwater habitats. This case study reveals that concern about freshwater resources on Sakhalin Island is not only about the water itself. Hundreds of streams and rivers flow across this island, providing drinking water and habitat for multiple species. Rather, concern centers on what is *in* the water, especially salmon, silt, and chemicals associated with hydrocarbon extraction. Local residents are concerned about pollution of fresh water by silt and chemicals only insomuch as the pollution affects the quality and quantity of salmon resources (Graybill, 2013). While concern about freshwater resources first emerged in the Soviet era with rapid urbanization, industrialization, and intense resource extraction, major concern did not emerge until the mid-1990s, continuing through the mid-2000s—a period characterized by the influx of foreign direct investment and TNC development of oil and gas deposits.

Alongside the demise of the campaign against freshwater degradation, local and indigeneous residents have received little social or economic gain (Graybill, 2008; Wilson, 2003), and ecologists note ecosystem-wide environmental degradation, especially during the development of infrastructure (e.g., construction of pipelines, roads) for Sakhalin's existing extraction projects. This degradation has affected in particular the bays and deltas, where both marine and riverine species have historically been rich in abundance and quality (Leonov et al., 2010; Polteva et al., 2009).

It is important to note that throughout the Russian Far East, but especially well documented on Sakhalin, indigenous and local peoples (re)turned to subsistence or semi-subsistence lifestyles after the collapse of the Soviet Union. On Sakhalin, this coincides roughly with the beginning of hydrocarbon development in the mid-1990s. With limited alternatives for income, many local and indigenous peoples have embraced subsistence salmon harvesting and overharvesting for the sale of salmon roe (Graybill, 2013; Newell, 2004). Along with industrial resource development, these practices have contributed to the decline of healthy salmon populations. Still, in many cases, local and indigenous people cite environmental pollution created by industrial resource development as the reason for depleted salmon resources today (Graybill, 2008).

In the Sea of Okhotsk region, the most contentious environmental transformations have occurred alongside offshore oil and gas development, affecting human and ecological communities on Sakhalin Island. Projections of the future suggest other parts of the Sea of Okhotsk will also suffer these effects, including the western shore of the Kamchatka Peninsula (Graybill, 2013). In many instances, local and indigenous peoples learn that their lands and waters are already being—or will soon be—developed for mineral and hydrocarbon resource extraction. Local communities fear that they will not have as much voice in socioeconomic impact assessments as they might if transnational companies were developing the resources (Sharakhmatova, 2011). Worried that salmon will disappear due to hydrocarbon extraction, inhabitants generally sense a lack of sustainable long-term development in the region, and so they consciously decide to contribute to salmon degradation in their own homelands (Graybill, 2009).

Concluding thoughts

The Russian Far East is a largely untapped frontier for both fish and hydrocarbon resources, both of which are accessed in and through water. The rush to extract multiple resources in this space leads to conflicts among the different actors (e.g., indigenous, local, TNCs) who use water for different purposes (e.g., subsistence fishing, industrial resource transport, industrial waste release). Additionally, water politics in Pacific Russia are not confined to freshwater spaces but extend to saline waters. While the marine region houses both salmon (when they live downstream) and offshore oil and gas (during upstream processing), both come onshore for different kinds of production at different stages in their life cycles: salmon migrate upstream to spawn, and hydrocarbons move inland for downstream processing. Thus, while the relatively open political and economic nature of post-Soviet Russia has created new gateways to Russia's resource peripheries, enticing TNCs interested in Russia's large and, in many cases, largely untapped hydrocarbon resources, conflict in this region has predominantly developed around concern about freshwater habitat and salmon health. This case study reveals that the potential for harm to fish and water resources by industrial resource development is high, creating the conditions for conflict between stakeholders dependent on fish and water resources for subsistence survival and stakeholders pursuing industrial-scale resource development.

Freshwater politics in this region are of concern to multiple stakeholders and largely relate to (1) changing quality of and access to salmon in freshwater systems experienced by local and indigenous subsistence resource users, (2) changing water quality in riverine and marine waters, of concern to human and non-human residents island-wide who depend on clean water, and (3) protection of newly (re)emergent cultural identities among local and indigenous communities with strong ties to the freshwater environment. While similar concerns about development and environmental degradation exist across Russia, the politics of environmental transformation resonate most poignantly where transnational development dominates socioeconomic transformation and where encroaching industrial development threaten conservation efforts, subsistence-level resource use, and cultural continuity.

Ultimately, the struggle over the health of Sakhalin's freshwater resources is most important to regional environments and the local and indigenous people who depend on them for survival. According to a Russian NGO, since 2006, strategies for political actors concerned about environmental and human wellbeing have changed: instead of fighting transnational hydrocarbon corporations in order to green extraction processes, those formerly siding with federal and regional governments against transnational development now find their targets to be the Russian government or global project financiers—entities that are far more difficult to target (Interview 13, 2015). Gazprom's entry into Sakhalin's development means changes in environmental political strategies, and environmentalists may have a murkier—and the environment a dirtier—future in this region.

Works cited

Augerot, X. (2005). *Atlas of Pacific salmon: The first map-based assessment of salmon in the North Pacific*. Berkeley, CA: University of California Press.

Bacheva, F., Kochladze, M., & Dennis, S. (2006, September). Boom time blues: Big oil's gender impacts in Azerbaijan, Georgia and Sakhalin. Report for CEE Bankwatch and Gender Action. Retrieved from http://genderaction.org/images/boomtimeblues.pdf

Bartels, D. & Bartels, A. (1995). *When the North was red: Aboriginal education in Soviet Russia*. Montreal: McGill-Queen's University Press.

Boentje, J. P. & Blinnikov, M.S. (2007). Post-Soviet forest fragmentation and loss in the green belt around Moscow, Russia (1991–2001): A remote sensing perspective. *Landscape and Urban Planning, 82, 4*, 208–221.

Boykevich, S. (2006, March 15). TNK-BP moves to solve Kovykta row. *Moscow Times*. Retrieved from http://www.themoscowtimes.com/sitemap/free/2006/3/article/tnk-bp-moves-to-solve-kovykta-row/206258.html

Bradshaw, M, (2005). 'Sakhalin's Oil and Gas Projects: Current Status and Future Prospects.' In Russia in Asia—Asia in Russia: Energy, Economics, and Regional Relations, Conference Proceedings. Washington D.C.: Kennan Institute, Woodrow Wilson International Center for Scholars, Occasional Paper 292, pp. 26–31

Bradshaw, M. (2006). Battle for Sakhalin. *The World Today, 62, 11,* 18–19.

Evans, A. B. (2012). Protests and civil society in Russia: The struggle for the Khimki Forest. *Communist and Post-Communist Studies, 45, 3–4,* 233–242.

Glavin, T. (2006). *The sixth extinction: Journeys among the lost and left behind.* New York: St Martin's Press.

Grant, B. (1995). *In the house of Soviet culture: A century of Perestroikas.* Princeton, NJ: Princeton University Press.

Graybill, J. K. (2007). Continuity and change: (Re)constructing geographies of the environment in late Soviet and post-Soviet Russia. *Area, 39, 1,* 6–19.

Graybill, J. K. (2008). Regional dimensions of Russia's transformation: Hydrocarbon development and emerging socio-environmental landscapes of Sakhalin Island. In Boren, T. (Ed.), *Kring Beringia: Expeditioner och folk. Geography and anthropology yearbook of the Swedish Society for Anthropology and Geography* no. 127, Stockholm, Sweden: Swedish Society for Anthropology and Geography.

Graybill, J. K. (2009). Places and identities on Sakhalin Island: Situating the emerging movements for "Sustainable Sakhalin." In Agyemen, J. & Ogneva- Himmelberger, Y. (Eds.), *Environmental justice and sustainability in the former Soviet Union.* Boston, MA: MIT Press.

Graybill, J. K. (2013). Mapping an emotional topography of an ecological homeland: The case of Sakhalin Island, Russia. *Emotion, Space and Society, 8,* 39–50.

Gustafson, T. (2013). *Wheel of fortune: The battle for oil and power in Russia.* Washington, DC: Thane Gustafson.

Heleniak, T. (1999). Out-migration and depopulation of the Russian North during the 1990s. *Post-Soviet Geography & Economics, 40, 3,* 155–205.

Henderson, S. (2011, March 18). Civil society in Russia: State society relations in the post-Yeltsin period. NCEEER Working Paper. Retrieved from http://www.ucis.pitt.edu/nceeer/2011_824–817_Henderson.pdf.

Henry, L. (2010). *Red to green: Environmental activism in post-Soviet Russia.* Ithaca, NY: Cornell University Press.

Interview 13. With Russian NGO, February 2015. Russia.

Ioffe, G. & Nefedova, T. (1998). *Continuity and change in rural Russia: A geographical perspective.* Boulder, CO: Westview Press.

Leonov, A. (1999). Oil and gas development on the Sakhalin Island shelf: An assessment of changes in the Okhotsk Sea ecosystem. Working paper of the Slavic Research Center, Sapporo, Japan: Hokkaido University. Retrieved from https://src-h.slav.hokudai.ac.jp/sakhalin/eng/71/leonov4.html.

Leonov, A. V. & Pishchal'nik, V. M. (2005). Analysis of conditions of petroleum hydrocarbon transformations in seawater and simulation of this process in Aniva Bay. *Water Resources, 32, 6,* 653–666.

Leonov, A. V., Pishchal'nik, V. M., (2006). Mathematical simulation of oil hydrocarbon decomposition in waters of Aniva Bay (Sakhalin Island, Russia). Proceedings of the 29th Arctic and Marine Oil spill Program Technical Seminar. Accessed online 1 August 2016: https://inis.iaea.org/search/search.aspx?orig_q=RN:37101854

Leonov, A. V., Pishchal'nik, V. M. & Chicherina, O. V., (2010). Transformations of biogenic substances in Tatar Strait water (the Sea of Japan): Analysis of mathematical modelling results. *Water Resources* 37(1): 33–57.

Meier, A. (2000, November 6). Arctic riches. *Time Europe, 156, 19,* 58–59.

Newell, J. (2004). *The Russian Far East: A reference guide for conservation and development.* McKinleyville, CA: Daniel & Daniel.

Polteva, A. V., Latkovskaya, E. M., & Leonov, A. V. (2009). Assessment of the background ecological state of water in the Chayvo Bay, Northeastern Sakhalin. *Water Resources, 36, 1,* 86–98.

Pryde, P. (1991). *Environmental management in the Soviet Union*. Cambridge, UK: Cambridge University Press.

Rand, P. S., Goslin, M., Gross, M. R., Irvine, J. R., Augerot, X., McHugh, P. A., & Bugaev, V. F. (2012). Global assessment of extinction risk to populations of sockeye salmon *Oncorhynchus nerka. PLoS ONE 7, 4,* e34065, doi:10.1371/journal.pone.0034065.

Rutledge, I. (2004). The Sakhalin II PSA–A production "non-sharing" agreement: Analysis of revenue distribution. Sheffield, UK: Sheffield Energy & Resources Information Services.

Sharakhmatova, V. (2011). Observations of climate change by Kamchatka indigenous peoples. Report for Lach Ethnoecological Information Center. Petropavlovsk-Kamchatsky, Russia: Kamchat Press.

Sinelschikova, Y. (2015). "No untouchables": Sakhalin governor arrested. *Russia beyond the Headlines.* Retrieved from: http://rbth.co.uk/society/2015/03/10/no_untouchables_sakhalin_governor_arrested_in_corruption_scandal_44343.html)

Slezkine, Y. (1996). *Arctic mirrors: Russia and the small peoples of the north.* Ithaca, NY Cornell University Press.

Stephan, J. (1971). *Sakhalin: A history.* London, UK: Clarendon Press.

Stephan, J. (1994). *The Russian Far East: A history*. Stanford, CA: Stanford University Press.

Thornton, J. & Ziegler, T. (2002). *Russia's Far East: A region at risk*. Seattle, WA: National Bureau of Asian Research and University of Washington Press.

Vakhtin, N. (1992). *Native peoples of the Russian Far North.* London, UK: Minority Rights Group.

Vysokov, M. (1996). *A brief history of Sakhalin and the Kurils.* Yuzhno-Sakhalinsk, Russia: Sakhalin Book Publishing House.

Weiner, D. (2000). *Models of nature: Ecology, conservation, and cultural revolution in Soviet Russia.* Pittsburgh, PA: University of Pittsburgh Press.

Williams, A. (2007). EBRD no longer considers current financing package for Sakhalin II. European Bank of Reconstruction and Development. Retrieved from http://www.ebrd.com

Wilson, E. (2003). Freedom and loss in a human landscape: Multinational oil exploitation and survival of reindeer herding in north-eastern Sakhalin, the Russian Far East. *Sibirica 3, 1,* 21–48.

Wood, A. & French, R. A. (1989). *The development of Siberia: People and resources.* Basingstoke Macmillan in association with the School of Slavonic and East European Studies. London: University of London.

Zhbanbova, P. (2013). The West Kamchatka Shelf. Paper presented at the North Pacific Regional Workshop to Facilitate the Description of Ecologically or Biologically Significant Marine Areas (EBSAs), Moscow. Retrieved from https://www.cbd.int/doc/meetings/mar/ebsa-np-01/other/ebsa-np-01-submission-wwf-russia-en.pdf.

Zimmermann, W. (2014). *Ruling Russia: Authoritarianism from the revolution to Putin.* Princeton, NJ: Princeton University Press.

Zonenshain, L. P., Kuzmin, M. I., & Natapov, L. M. (1990). *Geology of the USSR: A plate-tectonic synthesis*. Washington, DC: American Geophysical Union.

8 The plight of New Zealand's watersheds

Michael Vincent McGinnis

In *The Machine in the Garden*, Leo Marx (2000) describes two "kingdoms of force"—the pastoral ideal and the industrial modes of economic production and consumption. These two kingdoms of force represent a contested terrain of diverse epistemologies that shape human behavior and social identity. For Marx, pastoral heritage is a mode of consciousness—not simply a matter of expression but a matter of thought and behavior that shape relationships between human beings and nature. As an historian and literary critic of mechanistic society, Marx contends that modern experience is grounded in the tug-of-war between contrasting epistemologies of place: we long for a return to the past ecological state, while the impacts of techno-industrialism continue to transform the life-giving values of place. For Marx, human beings live in a "middle landscape" situated between pastoral heritage and a burgeoning mechanical area that represents two kingdoms of force that are actively shaping society and nature. Marx maintains that a "redemptive journey" is needed by members of industrial society to restore a more intimate relationship with the natural landscape and with place.

These two kingdoms of force have emerged in the contested terrain of watershed-based governance in New Zealand, or *Aotearoa* the *Māori* term for the 400 islands of the country and which translates as "the land of the long white cloud." Large-scale industrialization of the landscape and major watersheds threatens New Zealand's ecosystems and pastoral identity, and residents are torn between a desire to cling to their pastoral legacy and a desire to embrace economic growth and industrial development.

We can turn to the history of exploitation and colonization by Europeans to explain the paradox inherent to the New Zealand's green brand, 100% Pure NZ, and the plight of the country's watersheds. In *The Nomos of the Earth* (1950), German political theorist Carl Schmitt describes the origin of the Eurocentric global order that "discovered" the "New World." With respect to the Pacific Ocean, this "discovery" took place through colonization of lands long inhabited by indigenous peoples. Notable in Schmitt's discussion of the European epoch of global order is the role played by the ideals of economic development as a motivating factor in exploitation of other peoples' lands and resources. Schmitt argues that the notion of the "Old World as the center of the Earth" has been lost, and the industrial, technocentric worldview that became the arbiter in European and global politics has taken its place.

New Zealand is home to more than seventy major river systems (thirty on the North Island and forty on the South Island) as well as numerous streams and additional watercourses. Figure 8.1 depicts the major rivers of New Zealand. Of the combined total length of New Zealand's rivers and streams, 51 percent lies in catchments or watersheds with predominantly natural land cover such as native bush or alpine rock and tussock (Allibone et al., 2009). The remaining 49 percent of river length lies in modified catchments or watersheds (43 percent in agriculturally modified areas, five percent in plantation forestry, and one percent in urban settlement).

Consisting of 400 islands inhabited by both *Māori* and *Pākehā* (among other aboriginal and non-aboriginal peoples), New Zealand is a country shaped by race, class, and diverse social and cultural affiliations and identities. The *Māori* were the first to settle the islands approximately 850 years ago. In the early 1800s, British, European, and American whalers travelled to the islands across Oceania on a quest to harvest whale oil. In the *Māori* language, *Pākehā* refers to New Zealanders "of European descent." This is not a derogatory term but rather indicates the otherness of Europeans in the context of *Aotearoa*. During the nineteenth century, New Zealand and other islands were part of an expanding frontier available for the taking: fish, turtles, forests, and rich lands were harvested and traded. Whalers exchanged guns for *Māori* women and supplies, and these guns were then used by the fiercely war-making cultures of diverse *iwi* or tribes who fought over land and resources. In some cases, entire communities were wiped out in tribal conflicts well before Europeans arrived in New Zealand.

Māori and *Pākehā* are two kingdoms of force that embody alternative biocultural epistemologies, or worldviews, when it comes to the relationship between humans and nature. These contested social identities support diverse socio-ecological understandings of place and community. *Pākehā* is roughly translated as "one without breath." The traditional method of greeting in *Māori* is a *hongi* which involves pressing one's nose and forehead (at the same time) to another person at an initial encounter. While *hongi* serves a purpose similar to that of a formal handshake, the symbolic importance of it is much deeper, as the greeting resonates in the sharing of a common breath: in the *hongi*, the *ha* (breath of life) is exchanged. Through this exchange, greeters recognize that people are *of the land and the sea*. Each greeting reminds the greeters of their shared obligation to care for nature and community. This recognition of the connection between the landscape, the sea and the human community is deeply part of *Māori* culture.

This chapter explores two case studies. First, I describe the mythical value of the longfin eel (*Anguilla dieffenbachia*), or the *tuna*, in *Māori* culture. This species is endemic to the watersheds of New Zealand's two main islands, and its life cycle depends on both marine and watershed ecosystems. The use and protection of the eel is culturally significant to the *Māori*. Akin to wild salmon, the eel is a totem species that plays a role in *Māori* ceremony and custom. Ecologically speaking, the decline of the eel reveals the deterioration of the country's major

Figure 8.1 Major rivers of New Zealand

Source: Cartography by Gregory Woolston using data from Natural Earth, LINZ/New Zealand Hydrographic Authority, and National Institute of Water and Atmospheric Research, retrieved from https://www.niwa.co.nz.

watersheds. Ninety percent of pre-European wetlands have been drained, and around 70 percent of pre-European forest has been removed, radically reducing the native habitat of the eel.

A second case study examines an alternative epistemology associated with landscape, resource use, and the watershed commons. Within the last decade, New Zealand, which leads the world in dairy production, has provided roughly a third of the world's international dairy trade (DairyNZ, 2012). The country's landscape that is used to produce dairy, feeds more than 100 million people worldwide, and the dairy industry contributes 25 percent of New Zealand's merchandise export earnings (DairyNZ 2012). Roughly 37 percent of the country's land area is now devoted to dairy (DairyNZ 2012).

Land-use impacts of dairying in New Zealand illustrate dramatic change in the country's values when it comes to land and water. Growing dairy cattle populations contribute significantly to the decline of aquatic habitats and water quality as well as the loss of native species such as eel. Small lamb ranching operations have given way to multinational dairy ranching. Fonterra, globally and co-operatively owned by 13,000 New Zealanders who produce 22 billion liters of milk each year (DairyNZ, 2014), represents a great majority of the dairy production in New Zealand. Many people are reporting major changes in land use as well as in aquatic and marine systems as a result of growing dairy production. With changes in the landscape come changes in New Zealanders' social identity. Increasing dependence on the instrumental values associated with dairy production and the sacred, intrinsic values of the eel reflect epistemological conflicts and contested ecologies of social identity.

The fading sacred ecology of kaitiakitanga

Notwithstanding the impacts of the *Māori* people on the landscape, it is important to consider the *Māori*'s cultural knowledge, or what Fikret Berkes (1999) describes as the "sacred ecology" of indigenous peoples, in studying the natural history of rivers and watersheds. The *Māori* embrace a concept of *whakapapa*, or genealogy, which expands the concept of kinship to include the mountain, river, and coastal and marine areas they have traditionally inhabited. *Kaitiakitanga* is an essential part of *tangata whenua* (the *Māori* term for "people of the land") custom and practice (Kawharu, 1998; Roberts, Norman, Minhinnick, Wihongi, & Kirkwood, 1995; Ruru, 2009; Selby, Moore, & Mulholland, 2010). As described by Kahui and Richards (2014, p. 6), *kaitiakitanga* is "the embodiment of a nexus of beliefs geared towards survival, and provides an integrated management system that bridges the gap between ecology and people." *Māori* have passed the values of *whakapapa* and *kaitiakitanga* down through the generations. With these values the *atua*, or ancestors, have also passed down an obligation for stewardship.

As if to embrace and take in the breath of the river, *kaitiakitanga* is an ethic of place that reveals a deep connection between the mountain, river, and coastal area on the one hand and the human family, or tribe, on the other hand (Kawharu, 1998; Roberts et al., 1995). A broader sense of kinship in each tribe, or *iwi*, includes the more than human community. The epistemology of *kaitiakitanga* has inspired a range of conservation and restoration efforts throughout New Zealand (Roberts et al., 1995; Selby et al., 2010). Kahui and Richards (2014, p. 6) describe the sense of stewardship embraced by the Māori:

> This system is embodied in kaitiakitanga, which at a fundamental level incorporates a system of beliefs that pertains to the spiritual, environmental and human spheres. This provides the tools to manage the local environment; for example, spatial access and temporal access were regulated by *rahui* (reserved areas) and *owheo* (permanent conservation); maintenance of ecosystem structures were achieved by *ohu* (communal working bees) and rules pertaining to quantity and method of harvesting controlled extraction of resource units; metaphysical concepts such as *tapu* (involving restrictions of a spiritual nature) and *mauri* (vital essence; life force) enabled enforcement and monitoring of environmental indicators.

This multifaceted epistemology of *kaitiakitanga* was institutionalized in a number of treaties such as the Treaty of Waitangi (1840) as well as in related settlement statutes and laws established between diverse *Māori* and the British Crown, including the Resources Management Act of 1991 (Ruru, 2009). As a principle, *kaitiakitanga* is also importantly integrated into freshwater management and governance (Selby et al., 2010). The *Māori* sense of *kaitaikitanga* includes an intermingling of earth-based beliefs, communal practices, and egalitarian ethics that serve as a means to sustain resource use and place-based ecology (Kawharu, 1998). *Kaitaikitanga* emphasizes the link between ecosystems and human well-being and offers a more integrative approach to resource use, place, and community.

In New Zealand, the practice of *kaitiakitanga* remains a major source of cultural division and political turmoil because of diverse English and *Māori* interpretations (Waitangi Tribunal Report, 2012). Post-settlement treaties and conflicts over resource use, allocation, and reparations remain a major issue in the country. A sense of belonging to a place represented by cultural values such as *kaitiakitanga* broadens the sense of *iwi* kinship to *rivertime*, a sense that European settlers lacked. Rivertime is the cultural heritage that *iwi* are born into: it is an inherited right and obligation of stewardship (Selby et al., 2010). Use and access to resources is an important facet of tribal customs. *Māori* are major landowners: they own a portion of the land used for major industrial-scale agriculture. They also own major quotas in commercial fisheries (roughly 40 percent) and aquaculture (roughly 20 percent).

The plight of the longfin eel

The spiritual component of *kaitiakitanga* may simply become a fading reference point to a past practice. As *Māori* and *Pākehā* commercial enterprises transform nature into a mere commodity, *kaitiakitanga* and the river it celebrates falls under threat. The fate of the eel symbolizes changes taking place in New Zealand's watersheds. As a totem species for the *Māori*, the *tuna* is a galvanizing force that combines diverse elements of the natural and human worlds. In his poem, *A Tail for Maui's Wife*, Tuwhare (1974) describes the importance of *tuna*: "I am moved as water moved by eel's verve and impudence tuna: *I move with her I move against her I move inside her She is water.*" *Tuna* is a primary emblem in *Māori* mythology. James Prosek (2010) explains:

> [*Tuna*] is often found wrestling with Maui, the Pacific Islands' equivalent of Hercules. In several stories, diverse in their settings and tellings, Maui finds Tuna in bed with his wife, Hine, while she is sleeping. In other stories, a girl is cleaning her clothes in a spring-fed pool and is violated by an eel that was once her beloved pet. A villager captures the eel and cuts its head off, and the girl buries the eel in the sand. The first coconut tree then grows from it. In yet another story, Maui chops Tuna in half, the head becoming all the freshwater eels of the world and the tail all the saltwater eels of the world.

Believed to spawn at sea, possibly near the Cook Islands of the South Pacific, the longfin eel is currently listed as threatened by the New Zealand Department of Conservation (DOC) because of overfishing, hydrological modification, habitat loss, and pollution. In recent years, trap and transfer operations at some hydro-dam sites have revealed the number of longfin eels moving up the rivers to be very low—at least a 75 percent reduction—in stark contrast to the massive runs witnessed prior to the 1960s (Graynoth, 2006). Commercial catch records reveal a trend in decreasing size. Since the early 1990s, the commercial harvest of eels has halved due to rapid population decline.

Another key issue in the longfin eel's decline is habitat loss due to land-use changes and development, including wetland drainage, the construction of dams, irrigation schemes, river diversions, and culverts. Hydro-dams pose a particular problem, as they interfere with the eels' migration pathways. One estimate suggests that hydroelectric dams have blocked the longfin eel's access to the sea in 35 percent of its habitat (Graynoth, 2006). Some dams have special passes, allowing a small number of eels to bypass the massive concrete walls; however, unimpeded eel access is still crucial.

Watershed-based governance in New Zealand

Kaitaikitanga represents the long-held *Māori* value of linking human activities to the greater watershed ecosystem. Early *Māori* customs and practices are also reflected in the major policy reforms of the early 1990s. Scholars consider New Zealand's watershed-based program an ecologically progressive ideal for other countries to emulate (Ericksen, 1990). In the early 1990s, the country redrew its political boundaries to match the soft boundaries of each major watershed, a feat no other country has attempted to match. Regional and district councils assumed authority to manage both entire watersheds and marine areas extending twelve nautical miles into the ocean (i.e., the New Zealand territorial sea)

This watershed-based program emerged, in part, out of New Zealanders' historically strong identification with the landscape. In years past, small farming operations (e.g., wool and lamb agriculture) largely determined freshwater and land uses across New Zealand's two main islands. During World War II, the government guaranteed ranchers a set price for lamb, which was canned and distributed to soldiers fighting in Europe. With the price guarantee, lamb agriculture developed and expanded, and generations of sheep farmers have since passed down a pastoral heritage and place-based identity to their children. This pastoral identity was incorporated into the "green brand," 100% Pure New Zealand.

The Crown largely controls freshwater through the Resources Management Act of 1991 (RMA), an act that delegates authority to regional councils and whose passage was the fruit of policy and institutional innovation beginning in the late 1980s. The RMA's primary innovation is its recognition of the interconnections between watersheds and marine ecosystems, as well as its establishment of a governance framework for representing major catchment and watershed processes. Driven by a widespread desire to shrink the role of central government in planning and decision-making, New Zealand restructured its legal framework to emphasize regional authority (Ericksen, 1990). Under these reforms, more than 800 governmental and quasi-governmental agencies were dismantled or reorganized. In their place, three primary central government agencies and eighty-six local government authorities were established (twelve regional councils based on catchment boundaries and seventy-four territorial authorities called districts, or city councils). Collectively, these hold responsibility for all aspects of environmental, natural resource, and land-use planning and management (Ericksen, 1990). In addition, over fifty-five statutes and nineteen sets of regulations were eliminated and replaced by a single legislative enactment—the RMA—consolidating environment, natural resources, and land use beneath one umbrella for the purpose of promoting sustainable management of natural resources (Makgill & Rennie, 2012).

New Zealand created fifteen regional councils, reflecting the ecology of its major river systems. Regional councils oversee the taking, use, damming, diversion, and pollution of fresh water (Resources Management Act section 30 (1) (e) and (f)), and also have the power to set and decide on flow regimes as well as allocate water and activities in riverbeds. Councils exercise these powers through preparing regional policy statements (required) and regional plans (optional), and by their decisions on resource consent applications. Territorial authorities regulate activities on the surface of waters according to district rules.

The current regional governance framework manages human activities that cross the land and sea interface. Makgill and Rennie (2012, p. 149) note,

> [T]he definition of environment under the Resources Management Act (RMA) of 1991 includes ecosystems. That means that a proposed activity must also be considered in terms of its adverse effects on an ecosystem under section 5(2)(c). The ecosystem approach is also present in the division of regional councils' spheres of authority into catchments.

Yet, since RMA's passage, industrial-scale dairy farming has posed new threats. The failure of regional councils to address land-use changes associated with dairy farming is the subject to which I will now turn.

The rise of dairy production

Despite the intentions behind the RMA, major changes have occurred when it comes to sustaining the ecological integrity of New Zealand's watersheds. The current cow population has reached almost four million (DairyNZ, 2014). In 2007, 21 percent of dairies used rotaries, with two people milking up to 800 cows in two hours. New Zealand's dairy cow population is increasing faster than the resident human population. The shift to large-scale dairy production has also led to an overall increase in the density of cows, that is, the number of dairy cows per hectare, as well as in the amount of land owned and operated per farmer.

Dairy production not only raises concerns about water quality and the loss of aquatic habitat and associated native species diversity, it also raises concerns about water scarcity. Local water scarcity problems are manifesting across New Zealand due to of the overdrawing of rivers and creeks. Research on "virtual water" relationships between consumption, trade, and water resource use shows that freshwater resources should not merely be considered as an issue for individual countries or river basins. Exporting dairy and other animal products requires a significant amount of water, and the average water footprint per for one liter of milk is 1020 kg of water (Hoekstra, 2012). Overall, the total international virtual water flows related to the global trade of dairy and other animal products amount to 272 billion cubic meters per year—the equivalent of about half the Mississippi River's annual runoff in the United States (Mekonnen & Hoekstra, 2011).

Climate change is influencing the rainfall in New Zealand. As NBC News reported in 2013 (Newland, 2013), the country was facing its worst drought in thirty years. Parts of New Zealand's north island, where most dairy production takes place, were drier in 2013 than they had been in seventy years. Clearly, water scarcity is an economic reality in New Zealand, and the high amounts of water required to feed grass and dairy cattle exacerbates the challenge of adapting to climate disturbance and the likelihood of major droughts in the future. In 2013, the drought caused NZ$820 million in lost export earnings connected with dairy production. The drought also required ranchers to reduce the size of their dairy herds.

New Zealand is failing to live up to its brand of a clean, green country (Anderson, 2012). Many of the nation's watersheds have been converted to *wastesheds*: systems of shedding waste and other industry by-products (McGinnis, 2016). Scientists have documented the ecological impacts of the country's wastesheds, which are the subject of a heated political debate over the country's brand and future (Harding, 2007). As Mike Joy (2011), one of the country's outspoken freshwater scientists, observes:

> We have gone too far. Surely it is time to admit, even if just to ourselves, that far from being 100% pure, natural, clean, or even green, the real truth is we are an environmental/biodiversity catastrophe … In five decades, New Zealand has gone from a world-famous clean, green paradise to an ecologically compromised island nation near the bottom of the heap of so-called developed countries.

Watersheds and other aquatic habitats are essential to the maintenance of New Zealand's biodiversity. Many of the country's animals and plants are native species, that is, they can be found nowhere else. Specifically, over 80 percent of the country's 2,500 species of native conifers, flowering plants, and ferns are endemic (Rowan, Smith, Cochrane, Stephensen, & Gibbs, 1997). Of the estimated 245 species of birds that bred in New Zealand before human arrival, 71 percent were endemic, a result of the country's historic isolation from other land masses. Scientists estimate there to have been around 70,000 native land-based plant and animal species prior to settlement. From human settlement to 1994, 43 percent of the 93 endemic land, freshwater, and coastal bird species have become extinct, as have four of the 22 endemic seabird species (making 41% of all endemic species extinct) (Rowan et al., 1997). A 2009 study ranked 34 of 51 (or 67 percent) of native freshwater fish taxa as threatened or at risk (Allibone et al., 2009).

The effects of New Zealand's major shift to industrialized dairy have recently gained international attention. An analysis by Bradshaw, Giam, and Sodhi (2010) comparing loss of native vegetation, native habitat, number of endangered species, and water quality in 189 countries found New Zealand to be the eighteenth worst when it comes to preserving its ecosystems. These declines in fresh water health are, for the most part, related to agricultural impacts; excesses of sediment, phosphorous, and nitrogen, as well as faecal pathogens, continue to pose major threats to marine ecosystems.

The factors contributing to the decline of New Zealand's ecosystems are unmistakable. Nearly half the country's lakes and roughly 90 percent of its lowland rivers are now classified as significantly polluted. Nearly all parameters used to measure water quality reveal ongoing decline at most national and regional river monitoring sites. A dramatic decline in freshwater native species diversity also indicates ecological degradation. More than 60 percent of New Zealand's native freshwater fish, including the eel—as well as the only freshwater crayfish and mussel species listed—face the threat of extinction. Faecal contamination of the country's waterways is caused largely by dairy farming.

One reason for New Zealand's shift to industrial dairy is an increasing demand for milk products in Asia; New Zealand provides roughly four billion New Zealand dollars per year in dairy products to China (as of 2013) (DairyNZ, 2014). The demand for dairy in Asian markets is increasing as is the value of milk powder as an export. Due to public perception of Chinese milk as poor in quality, the Chinese have looked overseas for dairy and protein-rich animal products. In addition, with the loss of their primary protein sources from the sea from overfishing and pollution, Asian countries have shifted from fish to beef and dairy imports. As many of China's citizens gain greater wealth, the country's average daily intake of animal-source foods has more than tripled, which has led to an increased risk of heart disease, diabetes, and colon cancer.

New Zealand's Ministry for the Environment's issued a water report card in 2012 stating that water at 52 percent of monitored river sites is "poor" or "very poor" quality, while water at 28 percent of sites is "fair"—with a risk of illness for those swimming there. Only 20 percent of monitored river recreation sites received a grade of "good" or "very good".

Recent studies show that the decline in New Zealand's water quality is primarily due to dairy production (Morrison, Lowe, Parsons, Usmar, &McLeod, 2009). Trends in National Rivers Water Quality Network (NRWQN) data between 1989 and 2007 show an overall decline in water quality across the monitoring network, with significant increases in nitrogen (as nitrate and nitrite), dissolved reactive phosphorus, and total phosphorus levels. Numerous studies demonstrate the influence of land use on the quality of fresh water and show that diffuse agricultural pollution is the main cause of water quality degradation. In addition, water take has almost doubled in the past ten years. Freshwater surveys show that many places are not safe to swim.

While the New Zealand dairy industry exports dry milk powder to China, New Zealanders are increasing public opposition to dairy production. The country's pastoral heritage and the integrity of the environment are at stake. With the shift to dairy production, important land-use changes will likely contribute to the country's ecological insecurity as New Zealanders become more vulnerable to water shortages and a degraded landscape, which will in turn compromise their ability to grow food.

The future: the changing brand of 100% Pure New Zealand

With the rise of dairy cattle and the decline of eel and other species, the 100% Pure New Zealand brand faces challenges. The perceptions and values toward the landscape change as well. In this sense, the epistemology that shapes land use and watershed degradation is linked to ontology and identity. In the growing shadow of ecological degradation, New Zealand residents begin to question their own identities: the image of the pastoral Kiwi is fading as the landscape's industrialization becomes more apparent, documented, and discussed. Indeed, the fading brand is the subject of a contemporary debate among New Zealanders. New Zealand Prime Minister John Key compared the country's global 100% Pure New Zealand tourism marketing campaign to a fast food ad. "It's like saying 'McDonald's, I'm lovin' it'—I'm not sure every moment that someone's eating McDonald's, they're loving it ... It's the same thing with 100% Pure," he said. "It's got to be taken with a bit of a pinch of salt."

New Zealand's brand is a double-edged sword: even while New Zealand's economy may suffer the consequences of a brand that fails to live up to its promise, New Zealanders can also live up to the image that the brand provides.

Supporting this brand requires a deeper ecological approach to sustaining the social and economic foundations of healthy watersheds. Two distinct forces are at work in this brand—the pastoral heritage of the country and the mechanization or industrialization of the landscape.

New Zealand's industrialization threatens the brand and the tourism the brand celebrates. But the brand's foundation—New Zealand's wild country and pastoral heritage—is also threatened by changes in the social identities of the islands' diverse peoples. Ultimately, peeling back the brand's green veneer shows that industrialization of New Zealand's landscape has begun to erode the country's social and ecological image (McGinnis, 2012).

In less than a decade, the dairy cow has replaced the eel in New Zealand, and an epistemology of growth has supplanted the country's pastoral legacy. New Zealand's watershed-based approach to terrestrial, aquatic, and marine ecosystem issues is unique, but the country is not unique in its industrial pursuits and its participation in the global economy. We learn from the New Zealand experience that creating new political boundaries and institutions reflecting the soft boundaries of watershed ecosystems does not necessarily prevent the forces of global industrialization from taking hold and undermining earlier intentions. A watershed-based approach to governance is no panacea—it can only succeed when supported by the inherent values of healthy ecosystems and by those who act on behalf of those values. The political economy of a watershed changes with use and abuse; the impacts of industrialization and the overuse of freshwater ecosystems in New Zealand cannot be simply mitigated by redrawing administrative boundaries to "capture" the processes that shape and influence ecosystem integrity and health. The forceful socio-ecological impacts of industrial dairy production in New Zealand reflect a shift away from small-scale pastoral heritage, a break from the promise of watershed-based governance, and the eventual silencing of the cultural heritage of *kaitiakitanga*.

There remains another force at play that lies deep within the cultural heritage of the landscape and the sea—the historical epistemology and cultural affiliation with *Aotearoa* that has survived across many generations despite the colonialization by Europeans. The quest to recover the multiple values inherent to living *with* place is needed today in New Zealand, and this requires nothing less than a recovery of the commons and a return to the natural contract reinforced by *kaitiakitanga*. The plight of the eel is a symbol for the plight to live in accordance to *kaitiakitanga*. As the eel declines, so do the spiritual values associated with living in accordance to a watershed ideal. The *Māori* struggle to sustain *kaitiakitanga* as part of their social identity and cultural practice. The loss of a culture supportive of *kaitiakitanga* and the plight of the eel and watershed go hand in hand.

Many watershed-based activists in New Zealand are working to return the eel to the country's rivers. This restoration movement is analogous to the U.S. watershed movement insofar as activists are motivated by a sense of a lost place and an interest in re-inhabiting the landscape (McGinnis, 2016). The restoration of eel represents an attempt to restore a connection to place and the biotic community (Park, 1995). For decades, the inhabitants of New Zealand have been working to restore the eel (Graynoth, 2006; Selby et al., 2010). The eel brings together the *iwi* (and Kiwi), their respective places, home, and region: it connects the people to a common watershed that extends beyond the coast to the deep sea. In this sense, the eel embodies the irrevocable co-dependence between humans and New Zealand's terrestrial, river, and marine ecosystems along with the greater South Pacific. With the restoration of the eel to New Zealand rivers, this animal has once again become a primary source of food for the *iwi* and an essential facet of *iwi* ritual and cultural celebrations. The restoration of eel also traces a redemptive journey in addressing and responding to the industrial impacts of dairy production and other harsh land uses on the country's watersheds.

Acknowledgments

I would like to thank Charles Dawson, Teresa Shewry, Catherine Ashcraft, and Tamar Mayer for their useful insights on how to strengthen this chapter.

Works cited

Allibone, R., David, B., Hitchmough, R., Jellyman, D., Ling, N., Ravenscroft, P., & Waters, J. (2009). Conservation status of New Zealand freshwater fish. *New Zealand Journal of Marine and Freshwater Research, 44,* 271–287.

Anderson, C. (2012, November 16). New Zealand's green tourism push clashes with realities. *The New York Times.* Retrieved from http://www.nytimes.com/2012/11/17/business/global/new-zealands-green-tourism-push-clashes-with-realities.html

Berkes, F. (1999). *Sacred ecology: Traditional ecological knowledge and resource management.* Philadelphia, PA: Taylor and Francis.

Bradshaw, C. J. A., Giam, X., & Sodhi, N. S. (2010). Evaluating the relative environmental impact of countries. *PLoS ONE, 5, 5,* e10440, http://dx.doi.org/10.1371/journal.pone.0010440

DairyNZ (2012). *New Zealand dairy statistics 2011/2012.* Hamilton, New Zealand: DairyNZ.

DairyNZ (2014). *New Zealand dairy statistics 2013/2014.* Hamilton, New Zealand: DairyNZ.

Ericksen, N. F. (1990). New Zealand water planning and management: Evolution or revolution. In B. Mitchell, ed. *Integrated water management: International experiences and perspectives.* London: Belhaven Press.

Graynoth, E. (2006, June). The long and the short of it: Looking after the needs of native eels. *Details, 14,* 12. NIWA

Harding, R. (2007). Muddying the waters: Managing water quality in New Zealand. *Policy Quarterly, 3, 3,* 16–24.

Hoekstra, A. Y. (2012, April). The hidden water resource use behind meat and dairy. *Animal Frontiers, 2, 2,* 1–8.

Joy, M. (2011, April 25). The dying myth of a clean, green Aotearoa. *The New Zealand Herald.* Retrieved from http://www.nzherald.co.nz/business/news/article.cfm?c_id=3&objectid=10721337.

Kahui, V., & Richards, A. C. (2014). Lessons from resource management by indigenous Maori in New Zealand: Governing the ecosystems as a commons. *Ecological Economics, 102,* 1–7.

Kawharu, M. (1998). Dimensions of kaitiakitanga: An investigation of a customary Māori principle of resource management (unpublished doctoral dissertation). Oxford University, Oxford, UK.

Makgill, R. A., & Rennie, H. G. (2012). A model for integrated coastal management legislation: A principled analysis of New Zealand's Resource Management Act 1991. *International Journal of Marine and Coastal Law, 27,* 135–165.

Marx, L. (2000). *The machine in the garden: Technology and the pastoral idea in America* (35th anniversary edition). Oxford, UK: Oxford University Press.

McGinnis, M. V. (2012). Living up to the brand: Greening New Zealand's ocean policy. *Policy Quarterly, 8, 1,* 17–28.

McGinnis, M. V. (2016). *Science and sensibility: Negotiating an ecology of place.* Berkeley, CA: University of California Press.

Mekonnen, M. M., & Hoekstra, A. Y. (2011). National water footprint accounts: The green, blue and gray water footprint of production and consumption. Value of Water Resources Report Series Number 50. UNESCO-IHE, Delft, the Netherlands.

Ministry for the Environment. (2012). The natural resources sector briefing to incoming ministers. Wellington, NZ.

Morrison, M. A., Lowe, M. L., Parsons, D. M., Usmar, N. R., & McLeod, I. M. (2009). A review of land-based effects on coastal fisheries and supporting biodiversity in New Zealand. New Zealand Aquatic Environment and Biodiversity Report No. 37, Wellington, New Zealand.

Newland, J. (2013, March 15). New Zealand parched as worst drought in 30 years takes toll. NBC News. Retrieved from http://worldnews.nbcnews.com/_news/2013/03/15/17326669-new-zealand-parched-as-worst-drought-in-30-years-takes-toll?lite

Park, G. (1995). *Nga Uruora: Ecology and history in a New Zealand landscape.* Wellington, NZ: Victoria University Press.

Prosek, J. (2010, July/August). Maori Eels: New Zealand's Maori defend an extraordinary creature—and themselves. *Orion Magazine*. Available at: http://www.taiaoraukawa.co.nz/documents/Maori%20Eels%20-%20Taniwha%20James%20Prosek%2030%20June%202010.pdf

Roberts, M., Norman, W., Minhinnick, N., Wihongi, D., & Kirkwood, C. (1995). Kaitiakitanga: Maori perspectives on conservation. *Pacific Conservation Biology, 2,* 7–20.

Rowan, T., Smith, I., Cochrane, P., Stephensen, B., & Gibbs, N. (1997). *The state of New Zealand's environment 1997*. Wellington, NZ: Ministry for the Environment.

Ruru, J. (2009, October). *The legal voice of Māori in freshwater governance: A literature review.* Landcare Research, Lincoln, New Zealand.

Sahlins, M. (1976). *Culture and practical reason.* Chicago, IL: University of Chicago Press.

Schmitt, C. (2000). *The nomos of the earth in the international law of the Jus Publicum Europaeum.* (G. L. Ulmen, Trans.) New York: Telos Press. (Original work published 1950.)

Selby, R., Moore, P., & Mulholland, M. (2010). *Māori and the environment: Kaitiaki.* Wellington, NZ: Huia Books.

Tuwhare, H. (1974). *A tail for Maui's wife*. In *Something Nothing* (p. 43). Dunedin, NZ: Caveman Press.

Waitangi Tribunal Report (2012). The stage 1 report on the national freshwater and geothermal resources claim. Wai 2358., Wellington, New Zealand.

9 Cultural water wars

Power and hegemony in the semiotics of water

David Groenfeldt

Within the water sector, a war is being waged against nature. Rivers and aquifers are relentlessly over-exploited, and any rivers that seem to be in relatively better condition become targets for new dams, diversions, and inter-basin transfers. Even when riparian ecosystems collapse from too much water diverted, as has become the new normal for most river basins in the American West, proposals abound for yet more infrastructure to impound and divert still more water.

The war against nature, and specifically against water ecosystems, is also a cultural war of "hearts and minds" to gain adherents supporting the unfettered exploitation of water resources. How is it that indigenous and traditional cultural groups have sometimes, and indeed rather often, acquiesced in allowing the destruction of rivers they have traditionally revered? What were they thinking, and how might better understandings of that thought process help to develop moral inoculations to protect against further wanton destruction of the natural water ecosystems upon which we, and ironically, the future of our destructive civilization, ultimately depend?

The image I wish to convey is that of a cultural war embedded in a larger war of conquest against nature ("the West vs. the Rest," in the terminology of Marshall Sahlins, 1976, p. 54). My reading of global water development, and particularly the dramatic building boom of large dams since World War II, and which still continues, is that the concept of "war" is not only a metaphor; it is an accurate characterization of the mentality underlying large water infrastructure projects. However, the cultural war is not necessarily war-like. Indeed, quite the opposite can be the case. Cultural wars can be waged through gifts such as interest-free loans for constructing new dams or clearing the jungle to construct "high modern" irrigation schemes (Scott, 1999).

Political and even military power remain important, but, unlike in the past, when native populations could be slaughtered with impunity, as in the European conquest of America (Turner, 1980), conquerors today have more options for achieving hegemony without resorting to overt violence. Sometimes local cultures successfully resist both militarily and culturally, holding on to their traditional values and even strengthening their autonomy and sovereignty (Scott, 2009). But there is another possibility: the local population can peacefully convert to the conqueror's belief system, a prospect made all the more attractive by the same status asymmetry that makes it possible. Who wouldn't want to adopt the habits that have clearly brought success to the hegemonists?

The long global history of religious conversion, from the early Christians to ISIS, tells the story of gaining adherents, sometimes through brute force but often also through persuasion and example. The latter mechanism provides a frame for analyzing how diverse cultures around the world have come to adopt water policies and infrastructure design standards that are remarkably similar. What happened to the indigenous cultural values and beliefs about paying homage to sacred springs and rivers and respecting the inherent rights of all creatures to thrive in their natural habitats? Why are the dams, levees, and canals designed by indigenous engineers effectively identical to the designs of their Western counterparts? Does this prove that culture doesn't matter after all?

In this chapter, I argue that culture matters a lot, but the culture that matters most is that of the conquerors. Their favored status masks their tacit proselytization about how life should be lived, including how water should be managed. I believe we need to understand this "conquistador culture" and its quiet clash with traditional cultures if we are to find a cure for the ethical ills that plague the Anthropocene.

The case of Cochiti Dam, New Mexico

Why did the indigenous community of Cochiti in New Mexico (United States) agree to the construction of a mega-dam on the Rio Grande River, even though this waterway was, and still is, sacred to the community? As part of the deal, why did Cochiti's political leaders sell half the ancestral tribal lands for the development of a new city of 30,000 non-Indians overlooking the reservoir created by the dam? These fateful decisions were taken in the early 1960s by indigenous leaders who felt they had an obligation to transition their small, backward community into the modern era. The prospect of income from land development and recreational tourism related to the reservoir seemed like a good opportunity at the time (Lange, 1979).

Cochiti is one of nineteen indigenous communities still remaining in New Mexico after four and a half centuries of colonial occupation under Spain, Mexico, and, since 1848, the United States. The community's location on the banks of the Rio Grande River and immediately downstream from a major canyon, was identified by the U.S. Army Corps of Engineers (the Corps) as a promising spot for constructing a dam to protect the downstream city of Albuquerque from floods. Over a period of several years, the Corps convinced Cochiti's leaders that a dam would bring benefits: more secure irrigation supplies, tourism development from the recreational opportunities of the large reservoir surrounded by beautiful canyon walls, and a never-ending source of income from residential land development.

Constructing the largest (at the time) earth-filled dam in the world on lands belonging to an indigenous community for the purposes of controlling floods—which, we now know, were actually necessary for the ecological health of the river—met with little controversy in the 1950s and 1960s, when planning for this project was taking place. Cochiti tribal members were divided over the prospect of the dam, with recent war veterans convincing the more conservative tribal members that economic development, which the dam promised to spark, would usher in modern conveniences and opportunities. At the time, the economy of the village's 800 residents centered on small irrigated farms and traditional arts and crafts (jewelry, pottery, and drums). The dam would open doors for the community to participate in the mainstream American economic system.

Once the tribe conceded to the Corps' request, round-the-clock dam construction got underway, just two kilometers from what, for the previous 700 years, had been a peaceful village. Former Cochiti Governor, Regis Pecos, remembers:

> One of the most emotional periods in our history was watching our ancestors torn from their resting places, removed during excavation. The places of worship were dynamited, destroyed, and desecrated by the construction. The traditional homelands were destroyed. When the flood gates closed and waters filled Cochiti Lake, to see the devastation to all of the agricultural land upon which we had walked and had learned the lessons of life from our grandfathers destroyed before our eyes was like the world was coming to an end. And all we could do was watch.
>
> (Pecos, 2007)

The Corps' development scenario did not go according to plan. The land development corporation that had purchased rights to half the Pueblo's territory went bankrupt, leaving the land scarred with the bulldozed traces of unfinished residential streets. The Pueblo was able to buy back its land rights, but the dam remains. Downstream, the river's ecology has been severely affected: the wide, braided river channel, which had regularly flooded its banks, nourishing riparian cottonwood forests, morphed into a narrow, incised "hungry" river devoid of its customary load of sediment (now trapped behind the dam) and disconnected from the natural floodplain. Because of these changes in the river, the endangered Rio Grande silvery minnow faces extinction, while special forest restoration projects are needed to save the riparian cottonwood forests (Richard & Julien, 2003).

These impacts on the riparian ecology are not simply byproducts of ill-advised engineering but direct outcomes of a mindset that views the Rio Grande as a "floodway" rather than as a river. Environmental historian Daniel McCool (2012) refers to this attitude as "hubris." In building the dam, the Corps' engineers were quite literally waging a war on the Cochiti people's traditional worldview. The Cochiti cultural understanding of the river as a sacred, living being was being challenged by powerful forces of the colonial state, indeed, by the army of that state.

The nature of cultural water wars

Scholars and the popular media commonly raise the specter of "water wars" as a likely, and even inevitable, outcome of growing water scarcity. And though state-level armed conflicts over water remain hypothetical, various forms of institutionalized water conflict seem to be on the rise. "Water grabbing" takes place through state-promoted land sales, which include control over associated water resources. Water grabbing is also implemented through water privatization policies, which gained notoriety following the popular uprising in Cochabamba, Bolivia, in 2000 (Mehta, Veldwisch, & Franco, 2012).

While wresting water from local communities through physical or political means makes for good news headlines, cultural water wars are more subtle and, because they can so easily go unrecognized, are even more insidious. In cultural water wars, expropriating the meaning and values that local communities traditionally confer on water and water ecosystems becomes a primary weapon, and victims as well as perpetrators may be oblivious to the cultural destruction taking place. The war is recognized only after it is over, when traditional categories of meaning have been transformed into the dominant culture's terms.

Just as physical warfare has many variants, so does cultural warfare, and there is a wide diversity of ways and cultural contexts in which outside forces capture and co-opt traditional meanings attached to water ecosystems. What distinguishes true cultural warfare from political or physical warfare is that the vanquished "army" complies willingly with the cultural concepts of the conquering army. The battles of cultural warfare are psychological, unfolding inside the minds of both sides. Mander and Tauli-Corpuz (2006) refer to this type of conflict as "paradigm wars," but I prefer the term coined by Escobar (1995), "semiotic hegemony." This term connotes that not only is the traditional worldview being attacked, but even the cognitive building blocks for constructing that worldview are being challenged.

Semiotic hegemony

Semiotic hegemony is particularly insidious because it operates largely in the subconscious of both parties. The hegemonists typically see their role as teachers, advisors, or leaders, while their victims (for want of a better term) might see themselves as students or apprentices. Boelens and Vos (2012) cite the case of Peru's 2009 water law, which places high importance on water use efficiency, to the extent of issuing "efficiency certificates" to efficient users, qualifying them for priority in new water allocations. Agribusinesses, plantations, and even mines are considered "efficient" water users, and traditional peasant smallholders are encouraged to adopt similar practices. The former president of Peru, Alan Garciá, instructed his citizens to follow "the experience of successful peoples," which, he claimed was "the only way in which we can progress" (as cited in Boelens & Vos, 2007, p. 16).

Escobar (2008, p. 105) illustrates the concept of semiotic hegemony applied to conservation protected areas in indigenous territories. Conservationists unwittingly proselytize a rational, economic interpretation of nature: "Conservationist capital attempts a triple re-signification: of nature (as 'biodiversity reserve'), of local communities (as 'stewards of nature'), and of local knowledge (as knowledge of 'saving nature')....Once the semiotic conquest of nature is completed, the sustainable and rational use of the environment becomes an imperative."

While every situation is unique, we can distinguish between sudden "big bang" change, as in the case of Cochiti Pueblo, and gradual change, more typical of rural development. Both categories raise the issue of consultation and consent: To what extent were the target communities consulted? Was consent given? If so, what were the circumstances?

"Big bang" semiotic hegemony

The Cochiti Dam case illustrates a particular kind of "big bang" cultural warfare that can be termed a "Faustian bargain": Give up something very dear to you (in this case, the river's natural state and the Pueblo's ancestral lands, inundated by the reservoir), and I will offer you something so wonderful that you will have no regrets. Cochiti leaders had the legal option of saying no to the dam; they could have opted to continue their lives as before, but the dam proposal came at a time (soon after World War II) when change and the notion of progress were very much in the air. In the end, they did have regrets, partly because the scenario they had imagined did not go according to plan, and partly because their cultural conversion was incomplete. Traditional values related to land, water, nature, and their own community were still largely intact. The process of semiotic hegemony had been powerful enough to tilt the decision in favor of the dam, but just barely.

Semiotic hegemony is not necessarily permanent and can also catalyze cultural revitalization. While the Cochiti Dam was under construction, the Hopi Nation in northern Arizona agreed to a similarly Faustian project: a coal strip mine on their sacred Black Mesa with an open pipeline for transporting the coal as a slurry of coal and water. The source of water? A pristine glacial aquifer deep below the Hopi villages, the same aquifer that fed the springs Hopi people depended on in this arid climate. The Black Mesa mine opened in the mid-1960s and operated for some forty years, spurring political, environmental, and cultural protests that have arguably strengthened Hopi cultural identity (Groenfeldt, 2006; Whitely & Masayesva, 1998).

Semiotic hegemony does not always meet with easy success. The Orme Dam, proposed in the 1970s to be built on the Verde River near Phoenix, Arizona, would have inundated nearly two-thirds of Yavapai Indians' territory. Along with their land, their ancestors' graves would be lost, as well as the nesting area for several bald eagles. The Yavapai perceived that their entire culture was at stake. The tribe had suffered a history of hardships during nineteenth-century conflicts with the U.S. military; their lands had been reduced, but their rights to these remaining lands, on the Fort McDowell Reservation, were no longer in question. In a 1976 referendum, the tribal community voted against selling their land for the dam. The tribal chairman accepted their vote: "I have heard my people's answer. I don't care if you give a million dollars to each and everyone of us. Our answer will still be no" (as cited in Espeland, 1998, p. 208). November 12th, the day the U.S. secretary of the interior officially withdrew the dam proposal in 1981, has since become an occasion for celebration: the Orme Dam Victory Days, an annual event of the Fort McDowell Yavapai Nation.[1]

Gradual semiotic hegemony: rural development

The very concept of rural development contains a powerful component of semiotic hegemony, precisely the context Escobar (1995) had in mind when he introduced the term. "Development" implies that things could—and should—be different than they currently are. The purpose of development projects is to reform the status quo according to certain goals or standards originating from beyond the local cultural context.

In the early days of international development programs in India, the outside expert was invariably foreign, and often American. Arthur Mosher, an American agricultural economist and early proponent of the Green Revolution, sought to reframe the notion of farming as a business, with the farmer recast as an entrepreneur (Mosher, 1966). Today, India's agricultural sector is dominated by commercially-minded farmers who view the values of their grandparents as quaint but irrelevant for the modern era. The vast majority of India's freshwater resources is allocated to irrigation, supporting high-yielding crop varieties designed to thrive with the aid of commercial agro-inputs. No longer are foreign advisors the only actors promoting industrialized agriculture and massive water infrastructure. Indian experts, business leaders, and high-level politicians, including the prime minister himself are now the leading edge of semiotic hegemony (Debroy, Tellis, & Trevor, 2014).

Today, India is poised to embark on a major infrastructure project linking its major river basins and the rivers inside those basins (Debroy et al., 2014; Gupta & van der Zaag, 2008). High dams on the sacred Ganges River may seem out of place, but they are likely to be built in spite of objections from ecologists, social activists, spiritual leaders, and economists (Wu, Jeuland, Sadoff, & Whittington, 2013). The semiotic hegemony taking shape in India's water sector blends exogenous (e.g., World Bank) and endogenous forms of semiotic hegemony. Indeed, India's modern proclivity to massive engineering works is itself a coupling of outside (British colonial) hubris and homegrown hubris.

Vandana Shiva describes former Prime Minister Nehru's infatuation with dams as a part of the process of de-sanctifying nature. Nehru, India's first prime minister after independence, famously referred to large dams as "Temples of Modern India." Yet he changed his view in later life, confessing that he had suffered from the "disease of gigantism" (as cited in Shiva, 2002, p. 59). The Green Revolution, unfolding while India's initial large dams were under construction, had similar impacts. The complexity of the new agricultural input packages meant that control over the farming economy shifted from local communities to the state sector, and more recently to private corporations. Rather than encouraging the sort of local experimentation that has been the hallmark of so-called "traditional" farmers everywhere (see Richards, 1987), India's farmers were tricked into dependency and became addicted to the quick fix of technology (Shiva, 1992).

Embedded in Green Revolution technologies was the message that agriculture should focus on production and income goals. Farmers in India and throughout the Global South have taken this insidious teaching to heart, forsaking local agro-diversity in crops, seeds, and livestock in favor of monocrops relying on just one or two seed varieties. Yet at the same time, a growing cultural backlash to this kind of thinking is emerging in both developed and developing countries, favoring "multifunctional" approaches linked, in part, to local food traditions (Groenfeldt, 2007; Wilson, 2007). Both sides in this battle cite the urgency of meeting growing food demands in the context of a riskier climate. They are responding to the same world, but with very different worldviews.

Deconstructing semiotic hegemony

Unpacking the process whereby meaning is recast according to the terms of the dominant culture is the first step in arresting the process of semiotic hegemony, or at least pressing "pause" in order take time to consider what is going on. Cultural water wars and sub-wars (skirmishes) can revolve around almost anything—dams, rivers, institutions—because it is not the thing as a thing, but the thing as a repository and communicator of cultural meaning that has the capacity to motivate cultural conflict. In this section, we examine three broad categories of meaningful phenomena that give shape to a particular culture of water: 1) technical water features (e.g., canals and dams), 2) natural water features (e.g., a river), and 3) institutional features for managing and governing water.

Technical features: what is an irrigation canal?

In the Mahaweli irrigation project in Sri Lanka, a state-of-the-art United States Agency for International Development (USAID)-funded main canal was designed with steeply sloped concrete sides, making certain customary uses impossible such as bathing, laundry, and watering domestic animals. The project effectively re-defined an irrigation canal as having a single, technical purpose (conveying irrigation water) rather than multiple purposes (providing water for people and animals, as well as for crops). In the Bali Irrigation Improvement Project in Indonesia, funded by the Asian Development Bank, "improvements" were proposed to the traditional notched logs that divide the flow of a larger canal into several smaller field channels, with the amount of water in each channel proportional to the size of the area to be irrigated. The new project design called for replacing these notched logs with concrete "junction boxes." While both the traditional notched log and the modern junction box provided a technical solution for dividing a single large flow into calibrated smaller flows, the transparency of the traditional Balinese notched log also provided a way for farmers to visually monitor the proper functioning of the water division (Horst, 1996). With the new closed junction boxes, the actual dividing of the canal water took place invisibly inside the concrete box.

The new designs in Sri Lanka and Bali were introduced as the modern way of doing things, familiar to the authorities providing the funds, if not to the farmers cultivating the land. While beneficial from the perspective of technical efficiency, these new technologies shifted the relationship between the local people and the infrastructure on which they depended, communicating the message that water conveyance and management belong to a technical sphere of outside specialists. Neither religious offerings, nor social cooperation, nor the ancillary activities of washing clothes in the canal or children playing in the canal water have a place in the new culture of water that economic development now offers. Is this really a war? It looks peaceful on the surface, but cultural traditions and modes of meaning are most certainly under attack.

Natural features: what is a river?

Western science defines water as an inert mineral and rivers as resources to be managed. But this is not how most (and perhaps all) indigenous cultures view water. To the indigenous world, water is alive and conscious, and water ecosystems constitute part of an extended family of "all our relations." Among the indigenous peoples of Australia, these relationships imply both rights (e.g., to be acknowledged as a relative) and responsibilities (e.g., to protect the land, water, and associated plants and animals) (Jackson, 2006).

The ethics motivating indigenous peoples' use and governance of water are rooted in this sense of kinship with and responsibility for nature as a sacred duty and an ongoing spiritual practice. When the Corps labels the Rio Grande as a "floodway" rather than a river, the semiotic path is cleared to transform and dam the river that the Puebloan people of Cochiti had for countless generations viewed as sacred.

Institutions for water governance

Differences in cultural understandings about water take on tangible reality when it comes to rights of water access (Boelens, Bustamante, & de Vos, 2007). What constitutes fair allocation of water? Whose cultural understanding about water will be incorporated into the water governance arrangements? Such issues can be a source of conflict between indigenous communities and state water agencies and can also exacerbate internal social tensions within the indigenous community (van Koppen, 2007). For example, Hoogesteger (2015) shows that state and nongovernmental (NGO) technical assistance to a community irrigation system in Ecuador ended up undermining existing traditional governance institutions, resulting in internal conflicts.

The larger underlying conflict, however—and one that usually goes unnoticed—is the clash of worldviews, cultural schemas, and shared understandings about how water decisions should be made in the first place. Is water the purview of the state irrigation agency or of the local community? Is water held in common by a village, by individual farmers, or by each household? Local communities' answers to these questions may differ from the suggestions of outside actors. Too often, the manner of brokering these differences is tainted by cultural hegemony.

Beyond hegemony: how to bridge divergent water cultures

Just as physical wars involve a messy complex of partially aligned interests and infighting within warring factions, cultural wars are also messy affairs. From afar, the battle lines would appear to pit indigenous cultures against global neo-liberalism (Mander & Tauli-Corpuz, 2006). On closer inspection, however, the battle lines can be difficult to distinguish, as there are often conflicts within the indigenous community between factions supporting strict interpretation of traditional values and those embracing the economic opportunities of the proposed project. There are also cracks in the values of those imposing cultural domination. Conflicting views within the professional ranks of development experts are much in evidence on issues ranging from the rights of nature, collective economies, and climate-related alternative strategies for agriculture, to land and water management. How can we chart a course of cross-cultural water cooperation when there may be disputes even within the ranks of the warring parties? Here are a few suggestions:

Respect cultural distinctiveness and customary rights

In 2007, the United Nations Declaration on the Rights of Indigenous Peoples (UNDRIP) ratified the right of indigenous and traditional peoples to maintain cultural distinctiveness. The declaration, however, makes only two references to water. Article 25 states that "Indigenous peoples have the right to maintain and strengthen their distinctive spiritual relationship with their traditionally owned or otherwise occupied and used lands, territories, waters and coastal seas and other resources and to uphold their responsibilities to future generations in this regard." Article 32 addresses the delicate issue of control over natural resources and makes three points:

1 Indigenous peoples have the right to determine their own development priorities;
2 States shall consult and cooperate in good faith with the indigenous peoples concerned;
3 States shall provide effective mechanisms to mitigate adverse environmental, economic, social, cultural or spiritual impact (United Nations, 2008).

There is a wide gap, however, between the provisions of UNDRIP and contemporary legal frameworks at the national level. United States law still incorporates references to the Vatican's fifteenth-century "Doctrine of Discovery" as a justification for confiscating traditional indigenous territory (Miller, 2005, pp. 2–3). What can local traditional communities, and particularly indigenous communities, do to protect themselves against encroachments of the state? One strategy is diplomacy, where two sides enter into a negotiating process to find a mutually acceptable solution. However, as this chapter has argued, because of unequal power relations, coupled with the effects of semiotic hegemony, the final agreement is likely to favor the interests of the more powerful side.

A second strategy is to agree to disagree and aim for convivial separation. By framing water management as a political and ethical issue, and by claiming the basic rights outlined in the UNDRIP, indigenous peoples might create their own relationship to water without conforming to the categories of the dominant society (Palmer, 2006). A third strategy involves indigenous players embracing the mainstream legal "rules" (i.e., the Western legal framework) without embracing the values and ethics implicit in those rules. Thus, we find Maori attorneys arguing successfully in New Zealand that the Whanganui River should be accorded legal standing as a person. This legal fiction has a basis in both the Western legal framework and in Maori customary law; both legal systems recognize that "personhood" can be applied to entities, and not only to humans (see Groenfeldt, 2013, pp. 147–148).

Participation in water planning

Better planning is a commonly prescribed cure for the many shortcomings of water governance, but this approach is particularly challenging in a cross-cultural setting where cultures may hold fundamentally different ideas about water and water ecosystems. Even in situations where culturally distinct communities have the legal and political authority to conduct their own water planning and policy setting, as in the case of Cochiti Pueblo cited above, their ability to do so hinges on cultural clarity as well as technical capacity (Groenfeldt, 2003). Aside from Australia, which has engaged in a government-led effort to work with indigenous communities in developing water strategies (Jackson, 2009), very few comprehensive water resource management plans developed by, or even for, indigenous communities exist. Without a concerted effort to incorporate indigenous representation into water planning, development assistance projects are unlikely to represent indigenous interests.

Rethinking water quality

Many countries recognize the rights of their citizens to enjoy a healthy environment, including healthy rivers (Boyd, 2012), but what constitutes a reasonable standard when it comes to measuring a river's health? The right to be able to drink water directly from the river is an ultimate goal cited in many indigenous peoples' statements. A more limited aim cited in the 1972 U.S Environmental Protection Act is to make rivers "fishable and swimmable" (National Research Council, 2001) based on standards for "total maximum daily loads" (TMDLs) of specific contaminants.

Where a supportive legal framework exists, however, indigenous communities have additional options. Isleta Pueblo, a community of 2,500 located on the Rio Grande immediately downstream from the city of Albuquerque, New Mexico, leveraged the American legal system to force improvements in the river's water quality. Discharge from the city's wastewater treatment facility was in compliance with state and federal water standards, but the Pueblo set a higher water quality standard based on traditional religious practices which involved wading in the river and drinking small amounts of river water. The new standard required the city of Albuquerque to raise its effluent standards in order to comply with this religious purpose. As Dussias (1998) explains,

> By approving the ceremonial use designation, EPA, in effect, allowed the Pueblo to use its own understanding of the importance of water as a basis for establishing standards [and] is a great departure from the efforts of earlier federal government officials to eradicate the nature-based religious beliefs and practices of Native Americans and replace them with an understanding of natural resources as mere commodities readily available for exploitation.
> (pp. 659–660)

Conclusion: towards peaceful coexistence

When traditional views about water are at odds with the dominant society's beliefs and values, what happens? With few exceptions, the dominant society prevails. It is, after all, dominant, and can usually enforce the unstated principle, "might makes right." Cases where indigenous groups have prevailed against these odds, such as Isleta Pueblo's successful dispute with the city of Albuquerque, point to the importance of legal systems that recognize indigenous rights. But supportive legal systems are rare and are, in any case, subject to vagaries of enforcement and interpretation. The pressure of semiotic hegemony is always a danger where cultural power relations are highly skewed. The injustices suffered by indigenous peoples in the name of water development, such as forced displacement of entire communities and cultures, is now generally seen as a moral and ethical concern, and not simply an economic, legal, or political issue.

Perhaps the most critical element for constructive dialogue between politically unequal parties, for example traditional communities versus national water agencies, is to share and compare their conflicting values around water. In order to set the stage for dialogue, both sides need to make their values explicit. How do you regard the priority of healthy nature? Do you see this as a right, as a desirable feature, or as having no intrinsic value? Do you regard water as a fundamental commons, with everyone in the watershed functioning as an equal stakeholder? Or do you see water as a legitimate form of property which others cannot lay claim upon?

When indigenous and traditional communities can identify the values they hold dear, and when outside water agents—whether from the provincial water department, a national water department, or foreign development agency—can identify their own values and priorities around water and water use, the stage is set for constructive engagement. The intent should be to clarify the values of all stakeholders and arrive at a mutual understanding (not necessarily agreement) about the values that each side views as most important.

Once the values of all stakeholder groups are on the table, a consensus position on water values can start to take shape. The process is not simple, and it is likely to be messy, but success, here, is when both the local traditional representatives as well as outside experts successfully listen to each other and learn about the particular values motivating their counterparts. Ultimately, we are all human, and we need to adopt an approach that facilitates interchange at this basic human level. Cultural water wars do not need to continue indefinitely; a peaceful settlement is possible.

Note

1 http://www.fmyn.org/event/orme-dam-victory-days/

Works cited

Boelens, R., & Vos, J. (2012). The danger of naturalizing water policy concepts: Water productivity and efficiency discourses from field irrigation to virtual water trade. *Agricultural Water Management, 108*, 16–26.

Boelens, R., Bustamante, R., & de Vos, H. (2007). Legal pluralism and the politics of inclusion: Recognition and contestation of local water rights in the Andes. In B. van Koppen, M. Giordano, & J. Butterworth (Eds.), *Community-based water law and water resource management reform*. London: CAB International.

Boyd, D. (2012). The constitutional right to a healthy environment. *Environment: Science and Policy for Sustainable Development, 54*, 4, 3–15.

Debroy, B., Tellis, A. J., & Trevor, R. (Eds.). (2014). *Getting India back on track: An action agenda for reform*. Washington, DC: Carnegie Endowment for International Peace.

Dussias, A. M. (1998). Asserting a traditional environmental ethic: Recent developments in environmental regulation involving Native American tribes. *New England Law Review, 33*, 1, 1–653–1–667.

Escobar, A. (1995). *Encountering development: The making and unmaking of the Third World*. Princeton, NJ: Princeton University Press.

Escobar, A. (2008). *Territories of difference: Place, movements, life, redes*. Durham, NC: Duke University Press.

Espeland, W. N. (1998). *The struggle for water: Politics, rationality, and identity in the American Southwest*. Chicago, IL: University of Chicago Press.

Groenfeldt, D. (2003). The future of indigenous values: Cultural relativism in the face of economic development. *Futures, 35*, 917–929.

Groenfeldt, D. (2006). Water development and spiritual values in Western and indigenous societies. In M. Chibba, D. Nakashima, & R. Boelens (Eds.), *Water and indigenous peoples*. Paris: UNESCO.

Groenfeldt, D. (2007). How the multifunctionality concept can restore meaning to agriculture. In B. Haverkort & C. Reijntjes (Eds.), *Moving worldviews: Reshaping sciences, policies and practices for endogenous sustainable development*. Leusden, Netherlands: ETC/Compas.

Groenfeldt, D. (2013). *Water ethics: A values approach to solving the water crisis*. London: Routledge/Earthscan.

Gupta, J., & van der Zaag, P. (2008). Interbasin water transfers and integrated water resources management: Where engineering, science and politics interlock. *Physics and Chemistry of the Earth, 33*, 28–40.

Hoogesteger, J. (2015). Normative structures, collaboration and conflict in irrigation; a case study of the Pillaro North Canal Irrigation System, Ecuadorian Highlands. *International Journal of the Commons, 9*, 1, 398–415.

Horst, L. (1996). Intervention in irrigation water division in Bali, Indonesia: A case of farmers' circumvention of modern technology. In G. Diemer & F. Huibers (Eds.), *Crops, people and irrigation: Water allocation practices of farmers and engineers*. London: Intermediate Technology Publications.

Jackson, S. (2006). Compartmentalising culture: The articulation and consideration of indigenous values in water resource management. *Australian Geographer, 37*, 1, 19–31.

Jackson, S. (2009). National indigenous water planning forum: Background paper on Indigenous participation in water planning and access to water. Australia: CISRO.

Lange, C. H. (1979). Cochiti Pueblo. In W. Sturtevant & A. Ortiz (Eds.), *Handbook of North American Indians, Vol. 9, Southwest*. Washington, DC: Smithsonian Institution.

Mander, J., & Tauli-Corpuz, V. (Eds.). (2006). *Paradigm wars: Indigenous peoples' resistance to globalization*. San Francisco, CA: Sierra Club Books.

McCool, D. (2012). *River republic: The rise and fall of America's rivers*. New York: Columbia University Press.

Mehta, L., Veldwisch, G. J., & Franco, J. (2012). Introduction to the special issue: Water grabbing? Focus on the (re)appropriation of finite water resources. *Water Alternatives, 5*, 2, 193–207.

Miller, R. (2005). The doctrine of discovery in American Indian law. *Idaho Law Review, 42*, 1–122.

Mosher, A. T. (1966). *Getting agriculture moving: Essentials for development and modernization*. New York: Praeger.

National Research Council (2001). *Assessing the TMDL approach to water quality management*, Washington, DC: National Academy Press.

Palmer, L. (2006). "Nature", place and the recognition of indigenous polities. *Australian Geographer, 37*, 1, 33–43.

Pecos, R. (2007). The history of Cochiti Lake from the Pueblo perspective. *Natural Resources Journal, 47*, 3, 639–652.

Richard, G., & Julien, P. (2003). Dam impacts and restoration of an alluvial river—Rio Grande, New Mexico. *International Journal of Sediment Research, 18*, 2, 89–96.

Richards, P. (1987). *Indigenous agricultural revolution: Ecology and food production in West Africa*. Boulder, CO: Westview Press.

Scott, J. C. (1999). *Seeing like a state: How certain schemes to improve the human condition have failed*. New Haven, CT: Yale University Press.

Scott, J. C. (2009). *The art of not being governed: An anarchist history of Upland Southeast Asia*. New Haven, CT: Yale University Press.

Shiva, V. (1992). *The violence of the green revolution: Third World agriculture, ecology and politics*. London: Zed books.

Shiva, V. (2002). *Water wars: Privatization, pollution, and profit*. Cambridge, MA: South End Press.

Stanbury, P. (1996). The utility of tradition in Sri Lankan bureaucratic irrigation: The case of the Kirindi Oya project. In J. Mabry (Ed.), *Canals and communities: Small-scale irrigation systems*. Tucson, AZ: University of Arizona Press.

Turner, F. (1980). *Beyond geography: The Western spirit against the wilderness*. New York: Viking Press.

United Nations. (2008). United Nations Declaration on the Rights of Indigenous Peoples. New York: United Nations. Retrieved from http://www.un.org/esa/socdev/unpfii/documents/DRIPS_en.pdf

van Koppen, B. (2007). Dispossession at the interface of community-based water law and permit systems. In B. van Koppen, M. Giordano, & J. Butterworth (Eds.), *Community-based water law and water resource management reform*. London: CAB International.

Whitely, P., & Masayesva, V. (1998). The use and abuse of aquifers: Can the Hopi Indians survive multinational mining? In J. M. Danahue & B. R. Johnston (Eds.), *Water, culture and power: Local struggles in a global context*. Washington, DC: Island Press.

Wilson, G. A. (2007). *Multifunctional agriculture: A transition theory perspective*. Cambridge, MA: CABI.

Wu, X., Jeuland, M., Sadoff, C., & Whittington, D. (2013). Independence in water resource development in the Ganges: An economic analysis. *Water Policy, 15*, 3.

10 Theorizing gender, ethnic difference, and inequality in relation to water access and politics in southeastern Turkey

Leila M. Harris

This chapter makes two assertions. First, one cannot assess and fully understand the politics of fresh water without attention to inequality, notably with respect to gender and other axes of difference. Second, water access and politics often play a central role in constituting key categories of difference and inequality. As such, these categories are not static, but shift and change in relation to the changing waterscape and associated environmental dynamics. In this chapter, I elaborate these assertions with examples based on earlier work examining complex waterscape changes underway in the upper Tigris-Euphrates basin, also highlighting key concepts from several decades of work in feminist political ecology.

To begin, I detail several contextual features of the context in southeastern Turkey. Next, I support the chapter's two opening assertions through empirical discussion. In the final section of the chapter, I consider these insights in relation to broader themes in feminist political ecology (FPE), particularly taking into account conceptual and empirical linkages between socio-political difference, inequality, and freshwater politics. This final section aims to more fully draw out what an FPE approach—focusing on gender and other intersectional differences—offers for understanding connections between the focal themes of this volume—notably, water access, conflict, and identity. As elements of this work have been published previously, relevant selections are cited and can be consulted directly for further elaboration.

Context of the Southeastern Anatolia Project-related waterscape changes in southeastern Turkey

Southeastern Turkey is home to the headwaters of the Tigris and Euphrates Rivers—watercourses that hold historic importance for early Mesopotamian agricultural and urban settlements and that today provide key resources for agriculturalists and others throughout Turkey, Syria, and Iraq (Figure 10.1). Anyone familiar with this basin knows it has been, and remains, a site of intense politics and conflict—from several Iraq wars, to longstanding Kurdish separatism and state response, to recent violence associated with the rise of the "Islamic State." Turkey, the basin's upstream neighbor, generally enjoys the most economic and political stability of its co-riparians and has long focused attention on its southeastern border as the country's most impoverished region—the site of frequent violent clashes between the military and the Kurdish Workers Party (the PKK, a Kurdish separatism movement that operates with bases in the highlands of southeastern Anatolia and throughout the mountainous areas of bordering Syria and Iraq).

This region is the only one in Turkey dominated by the country's minority Kurdish-speaking population. (The southeast is also home to several other ethno-linguistic and religious minority populations, including Alevis, Arabs, Assyrians, Zazas, and Syriac Orthodox Christians). The southeastern region also stands out with respect to Turkey's long-term statist efforts to promote gender equity: basic statistics place natality in the southeastern Anatolia region at roughly double that of the country as a whole, while there is also high incidence of polygamy and domestic violence as well as relatively low rates of literacy among women.

This geography—of relative impoverishment, ongoing violent conflict, gender inequality, abundant water resources, and unequal power dynamics across the broader basin—sets the stage for GAP, the Southeastern Anatolia Project (or *Guneydoğu Anadolu Projesi* in Turkish). The project involves massive state-led development of the upper basin centered on hydroelectricity generation, water diversion for irrigation, and socio-economic and infrastructural development. Specific efforts include women's centers, literacy campaigns, and health initiatives to combat malaria and other diseases expected to accompany dam building, and ongoing conversion from rain-fed to irrigated agriculture (see Ünver, 1997 or Çarkoğlu & Eder, 1998 for an overview of the project).

Given the conflict in the region, one way to understand massive state investment in GAP is to read it as a developmental alternative to the considerable military investment of the past several decades. At present, considerable investment—and change—is underway, centered on large-scale transformation of the upper Tigris-Euphrates basin. With the present investment and large-scale transformation of the basin comes new livelihoods, economies, and agro-ecological realities and identities. It is to these changes—and their relationship to the two primary assertions of this chapter—that I now turn.

Ethnicity, gender, and socio-political differences: keys to understanding freshwater politics

Reading ethnicity: the Kurdish question and GAP development

Gender as well as other inequalities and socio-political differences matter to the freshwater politics of southeastern Turkey in manifold ways. First, from a geopolitical perspective and on a basin-wide scale, key inequalities between the countries are critical. Arguably, the GAP project could not move forward without relative inequalities between the basin countries. Turkey's geopolitical connections are far greater than its neighbors to the south and its situation upstream has been critical to the very ability to develop the river. (This position

Figure 10.1 Turkey's Southeastern Anatolia region, shown here in the dark color, is the focal region for development efforts as part of the GAP. As shown on the map, the GAP region is also upstream for the broader Tigris-Euphrates basin

Source: Cartography by Gregory Woolston using data from University of Minnesota Cartography Lab and Natural Earth.

of power relative to downstream riparians is sometimes referred to as hydro-hegemony; see Zeitoun & Warner, 2006.) This unilateral development is often cited as a reason for the high likelihood of future water-related conflict in the basin, considered by many as a "hot zone." Indeed, the GAP's damming and diversion of the Tigris and Euphrates Rivers is commonly understood as aggravating an already difficult situation, given that planned uses of the rivers require an estimated 150 percent of available fresh water (Hillel, 1994). Clearly, key inequalities between the countries are central to understanding current geopolitical and intra-basin tensions, the unfolding transformation of the waterscape, as well as the potential for future conflict.

While the literature primarily approaches the topic of water and conflict at the state (or country) scale, considering multiscalar dimensions of water and conflict, including the relationship of GAP water development to past and ongoing conflicts is essential (Harris, 2002). Consider, for instance, intra-state connections between waterscape changes and the Kurdish conflict that has long simmered in the region. Among the issues contributing to this conflict has been the question of cultural rights for Turkey's minority Kurdish population, including issues such as rights to school children in their native Kurdish language (a sizeable population estimated at nearly 20 million in a country of nearly 75 million). The GAP region has long been the central node of this conflict, with violent outbreaks. State forces have burned villages and forced the evacuation of rural communities. Under a long-term state of emergency in the region (lifted in the early 1990s), residents in the southeast were banned from associationalism or congregation—effectively making cultural celebrations and political protests impossible.

Highlighting these concerns foregrounds the politics, history, and ongoing conflict surrounding Kurdish identity as key lenses through which to understand contemporary fresh water-related changes and supports this article's first point—that ethnicity and other socio-political differences are key to understanding freshwater politics. To provide a few additional examples, an analysis of state-led GAP development reveals that, in part, the Kurdish conflict, and broader notions of "Kurdishness," are targeted through development of the twin rivers. Low literacy rates are often cited to convey the region's "underdevelopment": overall literacy in the GAP region is 60 percent compared with 80 percent in Turkey as a whole. (Note the significant gender disparity: 1990 figures suggest that in the southeastern region, 76 percent of men and nearly 45 percent of women are formally literate; GAP-RDA, 1998). Low rates of manufacturing and economic investment in the region along with related impoverishment also undoubtedly result (at least in part) from the longstanding conflict. These factors, combined with disruptions to local trading circuits, a concern aggravated by economic sanctions associated with the several Gulf Wars (see Yeğen, 1996), reduce companies' interest and ability to invest in the region. This complicated context invites consideration of the ways in which the politics—and the conflict—associated with the Kurdish question has set the stage for current GAP water-related changes.

Understanding linkages between illiteracy and Kurdish ethnic identity and politics (and the southeast's overall political economic situation) enables a different reading of the region's statistics. While arguing against literacy programs, particularly for women, may be difficult, contextualizing the situation is essential. Kurdish-language books, newspapers, and other print media were formally banned for several decades until the mid-1990s, Kurdish children were unable to attend primary school in their natal language, and long-term efforts to Turkify the population, including teaching Turkish, have characterized life throughout the region. Indeed, the freedom to teach, write, and even speak Kurdish was long an impetus for ongoing Kurdish resistance, and violence, with the prospect of jail time for those infractions (a situation that has arguably seen some improvement in the past decade).

In sum, when we consider indicators of "underdevelopment" cited to justify GAP damming and development, it becomes imperative to contextualize these "realities," taking into account contestations and politics related to Kurdish identity and rights. This history and context also offers a critical backdrop against which to understand the potential implications of GAP development, including how residents in the region might respond to changes (see Harris, 2008b and 2002 for discussion of the GAP project in relation to the Kurdish issue; see Dahlman, 2002 for details on the political geography of the southeast). Some claim that GAP changes may exacerbate political tensions in the southeast, with references to the "theft of Kurdish waters" or citing the privileging of certain ethnic groups or livelihoods over others (see Harris, 2002, 2006). More generally, while reception to literacy and other social development programs is also possible, and there has been some evidence that GAP-related changes and state interventions have been welcome for some in the region (Harris, 2012), the cultural and political overlay of such programs remains undeniable.

Adding this understanding of the cultural and historical context, we are able to understand that massive state investment in damming, water diversion, and social development programs in the region (approximately US$34 billion) is justified on a range of bases, from electricity needs to broader development goals. As well, features of Kurdish culture, including the *aşiret* social structure, are also frequently labeled as targets for intervention and modernization. Speaking to this, interviews with state planners related to the establishment of water user groups in newly irrigated areas suggest that emergent "democratic" organizations could counter longstanding hierarchical relations in the region, including the "tribal" *aşiret* structure as well as aspects of gender inequality (Harris, 2005).

In an interview, one planner commented on societal organization as a hurdle to progress: "The social structure in the region is problematic. Patriarchal, semi-feudal relationships are very dominant in the region, and this is an obstacle to development…Our social programs are primarily created to realize a vision of a participatory democratic individual, a modern individual." A government report expresses similar hopes that water user groups and irrigation-related changes will transform social relations in the region: "Irrigation can be regarded as a powerful tool for transforming the social structure and habits of the rural population in addition to bringing about a direct increase in their agricultural productivity" (Halcrow-Dolsar-RWC, 1994, p. 1, both cited in Harris, 2005, p. 186). These examples illustrate the ways in which ethnicity and social difference discursively and materially intertwine with freshwater politics in Turkey's contested border region.

Reading gender: gender difference, inequality, and GAP development

Idea(l)s related to gender frequently surface as aspects of social and economic disarticulation that also justify massive state intervention in the region—including elements where gender intersects with ethnicity, rurality, and livelihoods. (See Radcliffe & Westwood, 1996 for discussion of the ways in which indigenous women in the highlands of Ecuador are cast as "backward," and Mayer 2000 for broader discussion of linkages between gender and nationalism.) Recall, for instance, that among the important statistics used to justify GAP interventions are high natality rates (approximately double the national average), low literacy rates (particularly among women), and the ongoing practice of polygamy (Harris, 2008b). These gendered statistics and practices have become emblematic of southeastern Turkey's underdevelopment. Gender is also central to the developmental ideal of what southeastern Turkey should become: GAP development explicitly targets women's status, identities, and livelihoods in order to achieve "Western" or Kemalist, ideals. Women's centers, along with literacy and health initiatives, constitute perhaps the most visible of these efforts (Harris, 2008a; Harris & Atalan, 2002).

To historicize, and thereby better understand the focus on women as central to Turkish modernity and development, one only has to look at statements and policies dating to the country's first president and founding father, Mustafa Kemal Atatürk. Leading the new Republic in the 1920s and early 1930s, Atatürk often commented on the status of women, and famously suggested that it is impossible to allow one half of the population to improve and prosper, while ignoring the other half. Stemming from this position, the history of the Turkish Republic over the past ninety years has been associated with "state feminism." The state has highlighted the situation of women and implemented a number of policies ostensibly targeting gender inequality, including the high-profile and controversial headscarf ban, cited as key to achieving modernity (Saktanber, 2002; Harris, 2008b).

One straightforward point about the relationship between gender, inequality and freshwater changes relates to the differentiated effects on ongoing changes for various populations. In brief, difference and inequality are central to any reading of uneven distribution of and access to water resources. Consider, for instance, that with transition to irrigation (intended to be a major local benefit of the project), many residents of the pilot irrigation areas in the Harran Plain, particularly women, the landless poor, and those who heavily engaged in animal husbandry, have experienced notable losses in addition to gains. For many, expenses have increased, outpacing income gains. For some, livelihood options have suffered constraints (pasture is no longer available for grazing of animal herds), and work burdens have increased dramatically with the introduction of irrigation and related crop changes (significantly the predominance of cotton, now reaching 90 percent of the crop pattern of the Harran Plain).

Survey data show that men and women understand irrigation-related shifts differently (Harris & Karahan Kara, 2001). The data also reveal that landless people, those who are impoverished, and those who depend on animal husbandry for their livelihood also understand irrigated-related shifts differently. While the majority of residents experiencing shifts in irrigated agriculture view changes positively (nearly three-fourths of all respondents), the picture is not so simple. Analyzing the data in terms of population segments reveals important complexities: men tend to perceive the shifts more positively than women, well-off residents tend to perceive the shifts more positively than impoverished residents, and landed residents tend to perceive the shifts more positively than the landless. Those indicating the least positive associations with irrigation-related changes are self-identified poor women (only 33 percent responded positively) and landless women (44 percent had a positive response).

Another example of differing responses to irrigation changes among varying segments of the population surfaces in the area of work burden. All respondents recognized the increase in work burden that changes represent, yet different groups read these changes in different ways: 78 percent of middle-income men interpreted increasing work responsibilities as a positive change, compared with 36 percent of middle-income women. Clearly, important mechanisms regarding gender, poverty, landlessness, and ethnicity help explain differentiated experiences of irrigation-related changes for different segments of the population (Harris, 2008a).

Gender, ethnicity, and other categories of social and spatial difference decidedly underwrite Turkish state damming and diversion of the upper Tigris-Euphrates basin (Harris, 2008b). The examples above underscore the need to expressly consider gender, ethnicity, livelihoods, and other axes of difference as fundamental to understanding freshwater changes and politics—echoing and reinforcing broader themes in feminist political ecology (as highlighted in the concluding section).

Freshwater changes and politics as key to understanding the constitution and remaking of gender, ethnicity, and other differences/inequalities

The discussion that follows draws on a publication, 'Irrigation, gender, and social geographies of waterscape evolution in southeastern Turkey' (Harris, 2006), in which I employ Butlerian ideas of performativity (Butler, 1990, 1993) to suggest that gender and ethnicity matter not only for environmental changes (and diverse effects of changes, as with the above discussion), but that environmental changes, and specifically waterscape changes, matter for gender and ethnicity. In other words, it is in part through changing environmental and material conditions of the waterscape that gender, ethnicity, and other key notions of difference come to carry new and altered meaning and significance. Stated another way, as feminist political ecologists have long suggested, gender and other differences are central to conditioning resource access, management, and knowledges. Yet, here the discussion complements this analysis by also arguing, on the other hand, that environmental considerations and outcomes are also critical to conditioning operations and understandings of gender (and ethnicity as well as other categories of difference).

As I trace in this chapter, the GAP project has resulted in men and women renegotiating their labor practices, livelihoods, and subjectivities as a result of altered economies, crops, and changing environmental conditions. As the examples below illustrate, irrigated agricultural practices and differential water-resource geographies have become central to defining and consolidating the very terms of difference in the region—whether those differences have to do with masculinity and femininity, Kurdishness and Arabness, or landownership and landlessness. Once again, emphasizing an intersectional perspective is crucial (see the concluding section for a theoretical discussion of this concept). Speaking clearly to gender, but also linking to differences with respect to landholdings, ethnicity, and livelihoods, is of particular salience when attempting to understand ongoing social and political changes in the region.

Delivery of irrigation waters from the Euphrates River to the nearby Harran Plain, a pilot irrigation area near the border with Syria, has changed crop patterns, altered patterns of seasonal migration, and shifted labor expectations. Since the establishment of irrigation, growers have focused nearly exclusively on cash crops, with more than 90 percent of the cropping pattern eventually becoming dedicated to cotton. The movement to cash crops also involves a departure from the subsistence-based livelihoods that previously dominated the region, relying on rain-fed agriculture (primarily winter wheat, lentils, and other pulses, as well as livelihoods focused on the herding of sheep and goats). Now, with irrigation, there is a full-year growing season, focus on cash cropping, and, concurrently, less production of things that will be consumed directly by the household—instead cotton and other crops are now sold and food or other needs are then purchased on the market.

Regarding ethnicity, important shifts have been observed. Among them, narratives about the prioritization of Arabs over Kurdish farmers for water access, as well as shifting patterns of seasonal migration with new irrigated agriculture and associated seasonalities and cropping patterns. With respect to migration, it is important to note that, by and large, areas having yet to directly benefit from irrigation—areas from which seasonal migrants hail—were primarily to the north of the plain, including very near the Atatürk Dam reservoir, where the water originates. Meanwhile, the Harran Plain itself, to the southeast of the reservoir and just north of the Syrian border, is 80 percent Arabic speaking and 20 percent Kurdish speaking—making it an Arabic-speaking majority pocket within the broader Kurdish-dominated southeastern Anatolian region. The fact that areas remaining non-irrigated are predominantly Kurdish, while the initial irrigated areas are majority Arabic speaking is a coincidence of geography noted by some residents as evidence of inequities propagated by state agents. As expressed by a group of Kurdish seasonal migrant workers sitting around a fire after a day of work:

> They gave Arabs the water first, but not to us, we are Kurdish. This is not life. We will work to pick 100 *decars* of cotton. In Bozova, we also own 100 decars, but since there is no irrigation, our pistachio and nut trees don't give fruit soon after we plant them so we came here to get some money. We don't think we will earn good money here either, but we are obligated to come pick cotton.
>
> (Field notes Şanliurfa, 2001; Harris, 2006, p. 193)

As this statement illustrates, some Kurds feel they have been left out when it comes to the benefits associated with irrigated agriculture. These sentiments further reveal how changes in water-resource geography redefine meanings and associations attached to what it means to be Kurdish or Arab. Without oversimplifying the complex understandings and practices that constitute and maintain notions of ethnicity, in some ways Kurdish ethnicity is increasingly associated with seasonal workers (the numbers of Kurds in the Harran Plain swell during the harvest season), and Arab ethnicity is increasingly associated with farmers who employ seasonal workers and enjoy privileged, or more immediate, access to water. Consequently, although one of the principal aims of GAP development is to overcome longstanding disparities between this region and the rest of the country, the emergent waterscape also appears to be creating new intraregional disparities, reinforcing a differentiated social geography whereby some Kurdish residents feel their interests have, once again, been overlooked.

As these examples make clear, terms of difference are taking on new meaning and importance in relation to the new water-resource geography. How residents understand themselves and each other as Kurdish or Arab is now necessarily read through experiences of water-resource access, use, and change.

On a more positive note, improved livelihoods and economic well-being associated with GAP development could potentially attenuate longstanding conflicts and even help diffuse contentious state–society politics in the region. Lending force to this possibility, research has suggested that GAP-related changes are read by some as responding to concerns frequently raised by Kurdish separatists, including the need to foster political and economic development and inclusion of the region within Turkey on the whole. As such, there is evidence of heightened senses of state legitimacy as a function of irrigation access and other GAP services. For instance, interviews in the region discursively connect shifting perceptions of the state with GAP developmental changes: "At least the state turned its face towards us"; "The state thinks about us now"; or "We did not see any accomplishments of governments in the past we—we a little bit happy to see some now." Or as summed up by another respondent: "Our view of the state changed positively...We had hatred before, but now they started investing in the southeast" (all cited in Harris, 2009, pp. 10–11).

While responses varied—and there certainly were very negative responses as well—it is still important to consider the ways in which state irrigation delivery and other services (e.g. electricity or drinking water) may increase state legitimacy—a particularly important consideration given long-term contestation and violence in the region. As such, it is possible that ethnicity, gender, and state–society relations will be recast and redefined in relation to irrigation and other waterscape changes in important ways (Harris 2009, 2012).

With regard to gender, the agro-ecological shifts in the Harran Plain, particularly the shift from subsistence agriculture to the near mono-cropping of cotton, has brought marked changes in how men and women are viewed to be contributing to the household. With previous livelihoods there would be herding of sheep and goats (with direct access to meat, dairy, and wool), as well as the growing of wheat, lentils and other staple crops. With this subsistence economy, women made beds from wool, prepared cheese from milk, and processed meat for family food. Now, with cotton and cash cropping, men are negotiating the sale of cotton, and travelling to urban markets to purchase household goods. This severing of women's direct household contributions is at times narrated as: "Women no longer contribute to the household." With men who now purchasing items for the household, their status has increased, as this is interpreted as the head of household providing for his family's needs.

The discounting of women's contribution was obvious in several encounters I witnessed. An older farmer near the Syrian border stated, "Women in this village do not do any work. They don't go [to] the fields. They just make bread" (Interview, September 24, 2001). As he spoke, a group of us were sitting and drinking tea while watching a young bride and her mother-in-law make bread over an outdoor fire. The young bride, who overhead the comment, responded, "We also work hard. Is this not work? I'll also go [to the fields] to pick cotton tomorrow" (ibid.). Similar narratives were invoked repeatedly over the course of the nearly sixty interviews conducted throughout the Harran Plain, most often with husbands making claims that their wives do not work but "just sit." As well, women in rich families noted with pride that they do not work at all, only sit. Again, such erasures of women's domestic labor have the effect of enhancing men's status as providers for the household, as well as the status of those wives to do not have to work and can stay home with their children.

The arrival of irrigation affected men and women's relative status in other ways as well. Responding to the question, What has changed in your village since irrigation?, rural residents frequently cited changing material conditions, noting the escalating number of cars, cell phones, and cement houses, as well as the ability of richer men to purchase second and third wives—all visible status items reflecting the success of male heads of household.

Particularly noteworthy in terms of gender dimensions is the increasing incidence of cars throughout the plain. In this region, cars are driven exclusively by men, affording men greater mobility, scales of interaction, and access to resources than women. One Arab man noted that since irrigation, "Life has become better, people live in luxury, the city is so close to us now'" (Harris & Karahan Kara, 2001, survey A15). The respondent directly references the ability to travel more easily, making the city "closer" for men, but not necessarily for women, who remain in the village. This gendered mobility gap effectively widens gender differentials between men and women in ways that parallel my arguments about redefinitions of Arabs and Kurds. In effect, the newly irrigated landscape recasts the experiences and subjectivities of men and women. Masculinity and femininity take on new meaning, as men may increasingly frequent coffee shops and urban markets, enjoying higher status as providers, while women remain in the village to pick cotton, even as their labor contributions are erased and devalued.

Perhaps the most dramatic example of how gender is recast in relation to the emergent waterscape is illustrated by the work category of "irrigator." As widespread irrigation is recent in the plain, it is revealing that there is general agreement among residents of the plain that irrigation remains the domain of men, with over 90 percent of survey respondents noting irrigation as masculine labor. What explains the seemingly effortless and natural codification of irrigation work as male, rather than as shared or female work? When asked why women do not irrigate, respondents commonly cited the physical ardor (digging trenches requires considerable strength) as well as the schedule (irrigation operates on a 24-hour schedule, requiring farmers to work at night). If a woman were to engage in irrigation, her physical labor and unconventional working hours would call into question the integrity of her husband or male kin. The gendered nature of this work is so deeply embedded that even in cases where a woman is widowed, for instance, many insist that sons or even neighbors undertake irrigation on her behalf.

Several more stories illustrate the practices through which the difference between men and women with respect to irrigation is understood, cited, and made to appear as natural. As one Arab farmer explained, women do not irrigate because they cannot do so physically. He continued with pride, "At least among us Arabs you will not find any women undertaking irrigation. Perhaps among the Kurds you will find a few, but *our women* would not do that" (emphasis added, Interview, October 17, 2001). Another Arab farmer, Emre, explained that women do not irrigate because it is "too difficult." Several minutes after he made this statement, I asked Emre what he thought was the most difficult job, and he responded, "picking the cotton, but women and children do that" (Interview, October 19, 2001).

The idea of a naturalized difference between men and women in terms of the ability to participate in irrigation is brought into question by these two examples. In the first example, it is not only a question of women not physically being able to engage in irrigation labor, as sometimes (as the farmer admits) Kurdish women do irrigate. He assures us that "Arab" women do not do that. He is clearly invoking notions of appropriate femininity, in this case articulated in relation to understandings of ethnic difference. In fact, in over sixty interviews and during several months of observation, I encountered only one woman (indeed Kurdish) who actively irrigates, sometimes with her husband and at other times alone. Rather than emphasizing the difficulty of irrigation labor, she explains that since the arrival of irrigation, "life is easier—all I have to do is put the siphon in the irrigation canal and the field gets the water itself…It is easy" (Interview, September 28, 2001).

In the second example, Emre's idea that irrigation is "too difficult," and therefore not the domain of women, provides an interesting contrast to the work of picking cotton under the hot sun, which is dominated by women and children and is broadly considered the most difficult of agricultural tasks. In our survey, 40 percent of the respondents singled out cotton harvesting as the most difficult task, while only about 5 percent considered irrigation the hardest task.

Changing geographies and environments are key to understanding different labor practices, spaces of access, and influence, which in turn serve to cite and maintain sex-gender categories and other key social differences. Emergent irrigated waterscapes precipitate new labor, gender, and status geographies. Women increasingly move from work within and around the home to labor-intensive tasks in the cotton fields. With the shift from subsistence agriculture to cash cropping, men move from the village to urban networks and markets to purchase goods. The role of irrigator associated with the emergent agroeconomy is invoked as male and prestigious, while the role of harvester, conversely, invokes feminized labor, performed to a large degree by Kurdish migrant workers from the north (importantly intersecting gender with ethnic difference).

Masculinity and femininity, and the very notions of difference between men and women, are thus cited and sedimented in ways that reflect the new irrigation economy and ongoing waterscape changes. The categories of women and femininity take on a new meaning in terms of work obligations to the family as field laborers for the cotton harvest, while the categories of men and masculinity, similarly, take on a new meaning, and the masculine ideal becomes narrated as *he who is able to engage in prestigious work as irrigator, or who can control nature and manage water to produce crops and family income*. Articulated in relation to newly irrigated economies and ecologies, differences between male and female, or between Arabs and Kurds, are thus cited, maintained, and naturalized.

Conclusions: moving towards gender, FPE, and intersectional approaches

The two assertions that form the foundation of this chapter are only subtly different. The first, that one cannot assess and fully understand the politics of fresh water without attention to inequality, notably with respect to gender and other axes of difference, invites us to consider how differences matter for outcomes of the water-related changes. The second, that water access and politics often play a central role in constituting key categories of difference and inequality, invites us to consider how it is that the categories of Kurd or Arab, male and female are in part defined and understood in relation to water uses or management. Water uses (e.g. who irrigates) and management practices (who participates in management institutions; see Harris, 2005, 2006) are examples of the daily material practices through which the idea(l)s of sex difference and gender come to make sense and hold meaning.

Much has been written on the importance of gender, inequality, and difference in the contexts of environmental change and management, as well as the contexts of freshwater changes and politics. In particular, considerable attention has been given to the ways that men and women are situated differentially with respect to environmental and resource knowledges, conditions, access, and management (e.g. studies by Agarwal, 1988; Carney, 1993; Rocheleau, Thomas-Slayter, & Wangari, 1996; Sundberg, 2004). This chapter has highlighted these dynamics, while also emphasizing the importance of intersectionality in relation to poverty, ethnicity, and landlessness (cf. Harris 2008a; Nightingale, 2011).

As noted by Hawkins and Ojeda (2011), deciphering how to implement a truly intersectional approach remains a key challenge, though a strong imperative for current and ongoing work nonetheless. While FPE scholarship has increasingly adopted intersectional approaches, progress is needed in this area, including better theorization and greater attention to processes of racialization, postcoloniality, and other uneven power dynamics (Mollett & Faria, 2013).

In closing, this chapter has aimed to highlight several important ways in which gender, ethnicity, and other aspects of social difference and inequality are linked to freshwater access, politics, and conflict. I have argued that attention to difference and inequality is crucial to understandings of freshwater politics. As well, I have suggested that freshwater access and changes can reconfigure and reinscribe the very categories and meanings of social and political difference. This chapter is only one example of the ways in which feminist theory in general and feminist political ecology in particular can contribute to the study of water access and politics—pushing the boundaries of what is meant by the terms "feminist," "politics," and "ecology" (Harris, 2015. See also entire volume by Buechler & Hanson, 2015). Further application of feminist theory to freshwater politics is possible, and is likely to be fruitful for ongoing work. For instance, a recent essay by Neimanis (2012) highlights possibilities for novel politics as a feminist politics "responsively attuned to other watery bodies both human and more-than-human within global flows of power." As these examples highlight, FPE holds considerable potential for advancing our understanding of the complex linkages between identity, conflict, and freshwater politics.

Works cited

Agarwal, B. (1988). *Structures of patriarchy: State, community, and household in modernising Asia* (Vol. 2). London: Zed Books.

Buechler, S., & Hanson, A. (Eds.). (2015). *A political ecology of women, water, and global environmental change*. London: Routledge.

Butler, J. (1990). *Gender trouble: Feminism and the subversion of identity*. London and New York: Routledge.

Butler, J. (1993). *Bodies that matter: On the discursive limits of "sex."* New York: Routledge.

Çarkoglu, A., & Eder, M. (1998). The Southeast Anatolia Development Project. *cidc Insight: Turkey, No. 18*, 111–123.

Carney, J. (1993). Converting the wetlands, engendering the environment: The intersection of gender with agrarian change in the Gambia. *Economic Geography, 69*, 329–348.

Dahlman, C. (2002). The political geography of Kurdistan. *Eurasian Geography and Economics, 43*, 4, 271–299.

GAP-RDA. (1998). GAP Regional Development Administration. Ankara, Turkey.

Halcrow-Dolsar-RWC. (1994). Technical Discussion Paper No. 20: Assessment of Current Irrigation Design and Practices. GAP-RDA, Ankara Turkey.

Harris, L. (2002). Water and conflict geographies of the Southeastern Anatolia Project. *Society and Natural Resources, 15*, 8, 743–759.

Harris, L. M. (2005). Negotiating inequalities: Democracy, gender, and politics of difference in water user groups in southeastern Turkey. In F. Adaman & M. Arsel (Eds.), *Environmentalism in Turkey: Between democracy and development?* (pp. 185–200). Aldershot, UK: Ashgate.

Harris, L. M. (2006). Irrigation, gender, and social geographies of waterscape evolution in southeastern Turkey. *Environment and Planning D: Society and Space, 24*, 2, 187–213.

Harris, L. M. (2008a). Water rich, resource poor: Intersections of gender, poverty, and vulnerability in newly irrigated areas of southeastern Turkey. *World Development, 36*, 12, 2643–2662.

Harris, L. M. (2008b). Modernizing the nation: Postcolonialism, (post)development, and ambivalent spaces of difference in southeastern Turkey. *Geoforum, 39*, 51698–51708.

Harris, L. M. (2009). States at the limit: Tracing contemporary state-society relations in the borderlands of Southeastern Turkey. *European Journal of Turkish Studies [online], 10* Online at: http://ejts.revues.org/4122.

Harris, L. M. (2012). State as socio-natural effect: Variable and emergent geographies of the state in Southeastern Turkey. *Comparative Studies of South Asia, Africa, and the Middle East, 32*, 1, 25–39.

Harris, L. M. (2015). Foreword. In S. Buechler, & A. Hanson (Eds.), *A political ecology of women, water, and global environmental change* (pp. ixx–xxiii). London: Routledge.

Harris, L., & Karahan Kara, Z. (2001). *Survey of post-irrigation related changes in the Harran Plain*. Unpublished.

Harris, L. M., & Atalan, N. (2002). Developing women's spaces: Evaluation of the importance of sex-segregated spaces for gender and development goals in southeastern Turkey. *Kadin/Woman 2000, 3*, 2, 17–46.

Hawkins, R., & Ojeda, D. (2011). Gender and environment: Critical tradition and new challenges. *Environment and Planning D: Society and Space, 29*, 2, 237–253.

Hillel, D. (1994). *Rivers of Eden: The struggle for water and quest for peace in the Middle East*. New York: Oxford University.

Mayer, T. (2000) *Gender ironies of nationalism: Sexing the nation.* London: Routledge.

Mollett, S., & Faria, C. (2013). Messing with gender in feminist political ecology. *Geoforum, 45*, 116–125.

Neimanis, A. (2012). Feminist subjectivity, watered. *Feminist Review, 103,* 23–41.

Nightingale, A. (2011). Bounding difference: Intersectionality and the material production of gender, caste, class and environment in Nepal. *Geoforum, 42*, 2, 153–162.

Radcliffe, S., & Westwood, S. (1996). *Remaking the nation: Place, identity and politics in Latin America*. London: Routledge.

Rocheleau, D., Thomas-Slayter, B., & Wangari, E. (1996). *Feminist political ecology: Global issues and local experiences*. New York: Routledge.
Saktanber, A. (2002). *Living Islam: Women, religion and politicization of culture in Turkey*. London: IB Tauris.
Sundberg, J. (2004). Identities in the making: Conservation, gender and race in the Maya Biosphere Reserve, Guatemala. *Gender, Place and Culture, 11*, 1, 43–66.
Ünver, I. H. O. (1997). Southeastern Anatolia Project (GAP). *Water Resources Development, 13*, 4, 453–483.
Yegen, M. (1996). The Turkish state discourse and exclusion of Kurdish identity. In S. Kedourie (Ed.), *Turkey, identity, democracy politics* (pp. 216–229). London: Routledge.
Zeitoun, M., & Warner, J. (2006). Hydro-hegemony: A framework for analysis of transboundary water conflicts. *Water Policy, 8*, 435–460.

11 Water, life, and community in the arid realm

The curious case of *karez* and *acequia*

Daanish Mustafa

Introduction: lost water worlds under modernity

This contribution is a story of how two customary techno-social systems, *karez* in Central and West Asia and *acequia* in the southwestern United States, have mediated multiple societal expectations from water ranging from productivity and livelihoods to community and esthetics. *Karez* is an ancient system of underground water channels where water flows by gravity from the "motherwell" dug into the water table (Figure 11.1). The water channel flows underground for a distance of between several hundred meters and several kilometers before it emerges from the ground at the "daylight point." The underground channel has a series of wells for maintenance purposes, but water can be used for irrigation and other purposes only from the daylight point onwards. One of this system's main advantages is its reduction of otherwise high evaporation losses. The *karez* system, also called *qanat* in West Asia, *galleria* in Spain, and *foggara/khattera* in North Africa, is prevalent as far north as Central Asia, as far west as Morocco, and as far east as Japan. Some *karez*-like irrigation systems also operate in Latin America (Lightfoot, 1996a). Because the *karez* system depends on passive tapping of shallow groundwater, its water supply varies with seasonal changes in precipitation, and hence groundwater levels, and thus it does not pose a threat to the long-term sustainability of groundwater levels.

In this chapter, I will discuss case studies from Balochistan in Pakistan and the Nakhchivan region in Azerbaijan (Figures 11.2 and 11.3), as well as the comparable *acequia* system in the southwestern United States. In Balochistan, the *karez* system is under threat from modern tubewell pumps (van Steenbergen & Oliemans, 2002), just as it is in steady retreat in most of the other places where it is practiced (Beaumont, Bonin, & McLachlan, 1989; Lightfoot, 1996a, 1996b). In the Nakhchivan region of Azerbaijan, however, the *karez* system, which was extinct by the middle of the twentieth century because of excessive groundwater pumping, has recently been resurrected as a result of direct intervention by a United Nations project.

Before white Anglo settlers reached the southwestern United States in the nineteenth century, late sixteenth-century Spanish settlers had migrated north from Mexico and established the first permanent European settlement in what

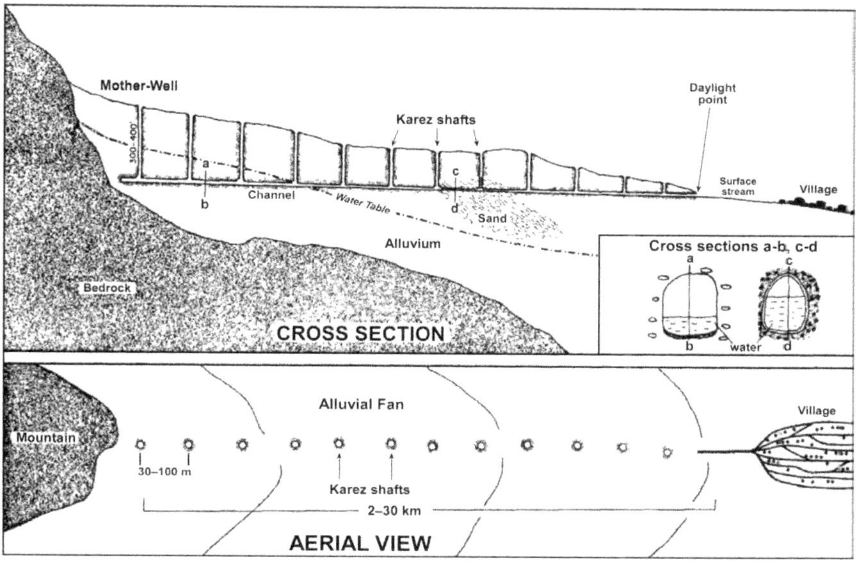

Figure 11.1 Schematic diagram of a *karez*

Source: Original by Professor Thomas F. Glick, Department of Geography, Boston University. Modified by the author and used with permission.

is now New Mexico. By virtue of their long historical association with Arab rule in Spain from 711 AD to its end in 1492, these Spaniards were well versed in Arab water management traditions. In the southwestern United States, they encountered the same climactic conditions within which many of the Arab water management techniques, especially of the *karez/qanat/foggara* system described above, had been practiced. In the southwestern United States, though, the new settlers dug surface irrigation ditches called *acequias* to transport water from stream channels to agricultural fields. Over the following two centuries, the Spanish/Hispanic settlers would build scores of new *acequias* premised not upon individual ownership of water, but rather on an ethos of communal ownership and management not so different from the Paiute and Pueblo Indian societies' view of water (Bates, Getches, MacDonnell, & Wilkinson, 1993; Brown & Ingram, 1987; Rodriguez, 2007). When Anglo settlers arrived in this region, *acequias* were the dominant water management structures they observed.

Acequia water law is a synthesis of Arabic, Spanish, Roman, and Pueblo Indian traditions (Hicks & Peña, 2003), and in this chapter, I will discuss the existing *acequia* communities in the southwestern United States in the context of the dominant individual property rights-based water management regime. In the process, I will compare the *karez* and *acequia* water communities to demonstrate the comparable wider set of values that the two systems try to satisfy beyond the basic livelihood and production functions of water.

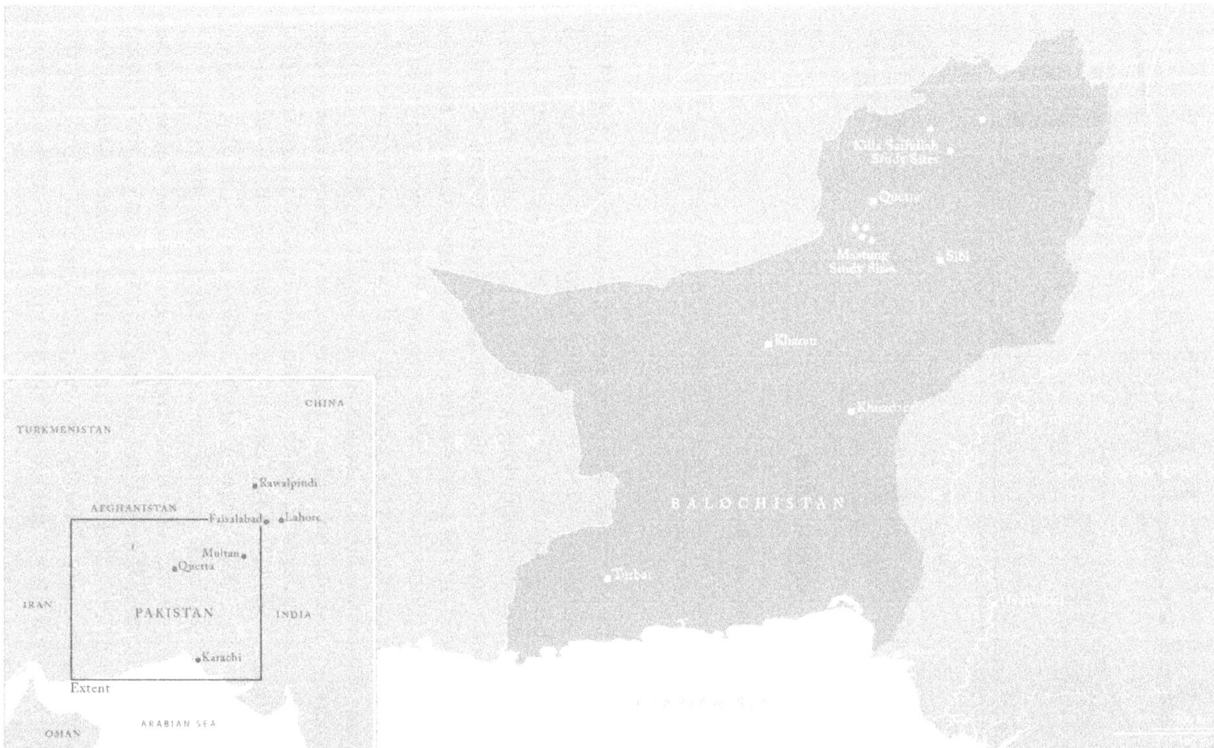

Figure 11.2 Balochistan Province in Pakistan. The black dots represent the communities where field research was undertaken by Mustafa and Qazi (2007)

Source: Cartography by Gregory Woolston using data from Natural Earth.

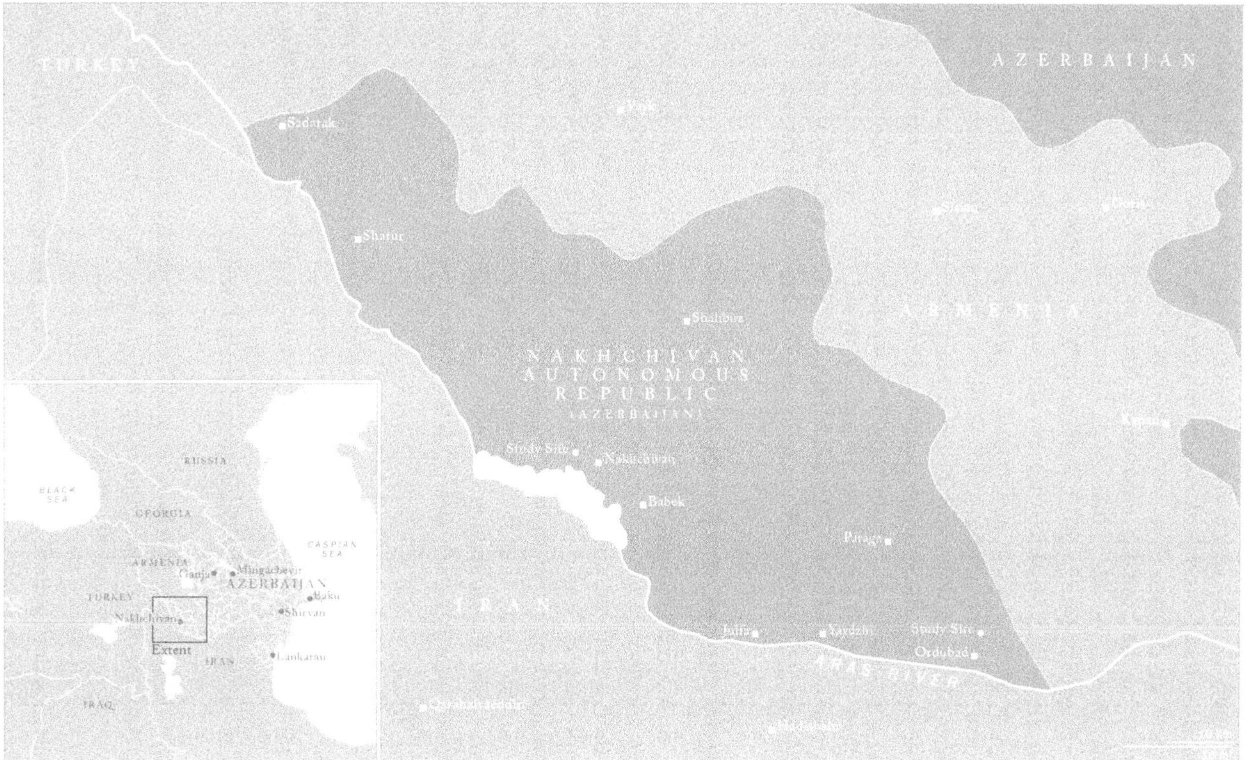

Figure 11.3 Nakhchivan Autonomous Republic in Azerbaijan

Source: Cartography by Gregory Woolston using data from Natural Earth.

There is an emergent interest in how water resources geographies are imbricated in power dynamics on local, national, and international scales (e.g., see Birkenholtz, 2009; Mustafa, 2002; Zeitoun et al., 2010). In a more political ecological mode, Erik Swyngedouw and his students have virtually pioneered research on how water has served as a conduit for state formation (Swyngedouw, 1999), citizenship (Loftus, 2007; Swyngedouw, 1997), and perpetuating power asymmetries and the capitalist metabolism (Bakker, 2003; Budds, 2004). This body of literature along with others (e.g., Mustafa & Reeder, 2009; Sneddon & Fox, 2006; Sultana & Budds, 2013; Zeitoun & Warner, 2006 to name a few) has focused on the political balance of water and how power asymmetries flowing from water politics produce and reproduce geographies of social power on local, national, and international scales. While this literature has legitimately and irretrievably politicized water and its geographies, there is a paucity of literature directly documenting the multiple worlds of meaning that water users attach to this slippery compound and showing how attention to these multiple values may yet yield practical knowledge in today's world (for a few exceptions to this charge, see Lahiri-Dutt, 2003; McMillin, McGinnis, and Woolson in this volume; Mustafa, 2013). This chapter engages that gap in the literature.

Karez, social capital, and power in Balochistan

For more than a millennium, the *karezes* of Balochistan not only served as a conduit for conveyance of water but also as the fulcrum on which water users balanced their economic, social, and cultural lives. The *karez* structures were communally owned. Water of the *karezes* flows perpetually, and very strong socially mediated property rights have emerged around the *karez* waters in addition to other interesting management features that point to the system's adaptive strength. The following quotations were gathered during field work in Balochistan between 2005–2007. A female water user from Balochistan said the following:

> Water was perpetual with a *karez* and there were fewer restrictions. We used to hear about a water-related meeting after years. [Since the demise of the *karez*] they have a water meeting almost monthly and sometimes weekly.

A male water user from Balochistan said the following:

> There is never conflict on *karez* water. There is conflict on communal land use and tubewell installation near the *karez*, e.g., the other tribe's tubewell is close to our *karez* and therefore there is conflict.

Balochistan is one of the few places in South Asia, perhaps even in the world, where, when a farmer is asked about his land holding (being a patriarchal society, the farmer is always a male), he would have no idea how much land he has. Land is practically unlimited in Balochistan, and, therefore, holding more or less of it is virtually irrelevant. The key limitation to farming is not land, but water, and therefore rural hierarchy is determined not by variation in land holding size but by the size of the water right one traditionally held in a *karez*. Water rights in a *karez* are divided into twelve to thirty-two day cycles, locally known as *shabanas*. The *shabanas* are further subdivided into three-hour increments locally known as *pass*. The perpetually flowing *karez* water is distributed among the *karez* shareholders on a fixed-time rotational system, and the water rights documented by Mustafa (2013) in Balochistan range from an hour to a few days. These well known and recognized rights are tradable among water users, almost functioning as a currency in village communities. Water rights are hereditary and are typically proportional to the original water purchase or contribution of families to the construction of the *karez* infrastructure. Since water rights in a given community are well established, conflict over *karez* water is virtually unknown. Furthermore, because everybody owns two parcels of land in a *karez* command area, one at the head of the *karez* and the other at its tail, the common problem of conflict between head and tail is also avoided.

In a *karez* community, a *mir-e-aab* serves as water manager and is responsible for administering water rights as well as mobilizing labor and resources for routine infrastructure maintenance. The officeholder is known as *hisabgar* in Azerbaijan. The *mir-e-aab* is typically a respectable male community member, and occasionally also a community leader. This position is comparable to that of the *mayordomo* in the *acequia* communities of the western United States, discussed later in this chapter. In many *karez* communities, up to four *shabanas* are held in communal reserve and are auctioned at the beginning of every cropping season to raise resources for infrastructure maintenance.

A number of rules protect the integrity of the *karez*, including the *harim* rule whereby, in Balochistan, no new *karez* motherwell, dug well, or water pump may be installed within 1200 feet of an existing *karez* motherwell. One of the quotes above mentions conflict around violation of such a *harim* rule. Furthermore, any activity that may adversely affect a *karez's* water quantity or quality is not permissible within the same distance of the motherwell.

The expansion of the electric grid to far corners of Balochistan beginning in the 1970s was part of state-sponsored efforts to modernize the province's agricultural sector. This modernization manifested itself in a proliferation of tubewells (electric and diesel water pumps) and promotion of water-intensive cash crops such as apples, onions, and coriander. Tubewells were promoted in particular because *karez* irrigation was deemed antiquated and wasteful, particularly on account of its perpetual water flow and because of seepage losses and seasonal variations in water flow (Kahlown & Hamilton, 1994). Tubewells were meant to help transition the region from supply-based irrigation to demand-based irrigation.

Thanks to flat rate electricity charges on tubewells and major subsidies on their installation (sometimes amounting to the full cost of installation), the number of officially recorded tubewells rose from around 2000 in 1970/71 to more than 14,000 in 2001/02. To this should be added the substantial number of private electric tubewells without a legal connection, as well as more than 14,000 diesel-driven wells in valleys not at the time connected to the grid or serving water users unable to afford the capital expense of electric tubewells (Qureshi, Shah, & Akhter, 2003). These tubewells were part of the state's attempts to assert itself through water development (e.g., see Swyngedouw, 1999).

The severe drought of 1998–2003 further diminished groundwater levels, which had been rapidly declining at a rate of 0.25–1.10 m/annum as early as 1992 (Water and Power Development Authority, 1992). The stressed aquifers came under immense pressure due to an acute absence of recharge, leading to sharp reductions in flow and, subsequently, the drying up of *karezes*. To cope, wealthy farmers installed new tubewells and deepened existing ones, while smaller farmers who could not afford to invest in groundwater extraction migrated to other places in search of alternate sources of livelihood.

In many cases, this withdrawal of *karez* shareholders caused the social organization around this traditional method of irrigation to disintegrate. *Karez* irrigation requires mobilization of considerable amounts of human and physical capital for purposes of maintenance and operation, and the mobilization of such capital in turn forms robust social capital among *karez* water users. Every water user within a *karez* is expected to contribute to maintenance proportional to his, and it is invariably a male, share in the *karez*. The *karez* decline, then, did not just mean decline in water property rights among *karez* owners but also in the social capital that had gelled around the technique (Mustafa, 2013).

Among other consequences, the *karez* decline has enhanced larger farmers' power and contributed to the dispossession and pauperization of smaller farmers, as the following quotes from water users illustrate:

> The small shareholders have really lost out because they do not have the resources to install their own tubewells. They are now forced into sharecropping arrangements with the tubewell owners.
>
> (Female water user, Balochistan)

> The rich have grown richer and the poor have grown poorer. The rich have installed tubewells and saved their orchards and crops. The poor farmer, on the other hand, does not have the wherewithal to save his few trees.
>
> (Female water user, Balochistan)

Beyond the loss of livelihood among many of the poor, there are other losses, including the loss of communal life and the world it created, as well as rising anxiety about the new world being ushered in through the introduction of tubewell irrigation, as the following quotes illustrate:

> In these modern times, people are abandoning the ancient and traditional irrigation, and trying to maximize cropped area. Ever since the tubewells have arrived, a competitive trend has emerged amongst the people and farmers. The installation of tubewells for modern irrigation succeeded in

> increasing agricultural productivity, but it also gravely damaged the ancient *karez* system. *Karezes* were a great source of social and communal life for us village folks. People would sit on their sides and discuss their issues and find solutions to their problems.
>
> (Ghaus Bux, Kunghar *Karez*, Balochistan)

> Our elders tell us that in olden times there was so much water in the *karez* that people working far away in the fields would receive their mid-day meals through the *karez*. The women folk would put food in a buoyant sealed container called "Danko," and put it in the *karez* near them. The *karez* would then take the containers to the farmers working downstream. Now those were the days! Nowadays times have changed. New technology is coming in, with new machines, and new people who like the latest machines. The traditional irrigation is progressively being forgotten.
>
> (Khaliqdad, Kunghar *Karez*, Balochistan)

The *karez* was a focal point for the social lives of women and children (see Figure 11.4). Many women often reported the *karez* as a preferred place where they could interact with each other in addition to conducting household chores such as bathing, laundry, and dishwashing. The loss of this aspect of community life coincided with the emergence of an insidious, violent strain in the culture of rural communities. Young people released from tightly knit bonds of rural communities, once relocated in alienating urban environments, or having observed pauperization of their own families, reverted to violent extremist

Figure 11.4 Children doing laundry and swimming in *Karez* Kunghar in Balochistan.

Source: Author.

organizations in Balochistan or espoused jihad in neighboring Afghanistan or against religious minorities and the state in Pakistan.

In popular parlance, modernity and technology are equated with competitiveness, productivity, and individual profit, while tradition, and its avatar, the *karez*, is equated with community, inclusion, and esthetics. The Pakistani state sees itself as a purveyor of modernity and concomitant nationalism, viewing tradition as primordial and anti-national. The consequences of ignoring the *karez* system's multiple values, amount to losses far beyond that of livelihoods. This chapter argues that these less obvious values are no less important in the lives of water users than the more obvious ones—as evidenced by the resurrection of the *karez* system in Azerbaijan.

The Phoenix-like life of (Kahrez) *Karez* in Nakhchivan, Azerbaijan

The *karez* system in Azerbaijan virtually went extinct during the Soviet era (1925–1990), partially because of the adoption of electric tubewells (locally known as artesians) as a preferred technology for pumping groundwater, and partially because of social upheaval caused by collectivization and nationalization. As the tubewells operating on subsidized electricity became communities' main source of water, the water table declined, as did motivation to maintain the *karezes*. Quite possibly, the absence of an individual stake in the *karez* also contributed to their decline.

The virtually extinct infrastructure was being resurrected by an International Organization for Migration (IOM)-managed Swiss government-funded project. The program aimed to strengthen place-based livelihoods in local communities and was using the *karez* as a conduit for accomplishing this wider objective. The project had initially focused on the civil works and infrastructural improvements of the *karez* tunnels and channels starting in 2004. But, as the project progressed through the late 2000s, it became clear that the social/institutional aspects of *karez* are much more critical than initially thought.

Pre-Soviet *karez* management systems closely paralleled water distribution regimes and institutions prevalent in many *karez* systems in other parts of the world, including in Balochistan as detailed above. With the rehabilitation of *karezes*, many communities are enacting water distribution regimes based on a fortnightly, fixed-time rotational water distribution schedule, where individual shares are proportional to the amount of land owned by individual community members. During the Soviet rule almost everybody had equal parcels of land, therefore since the end of it, the water rights attached to the land are also generally equal. In other *karezes*, however, communities are dividing up water in proportion to community members' contributions to the *karez* rehabilitation fund rather than by land ownership.

In some communities, this water distribution system resurrected in modern times is an attempt at balancing the customary water distribution regime with the modern reality of newer members and a post-Soviet land ownership pattern

(Mustafa, 2013). The point is that the customary water distribution system has not gone completely extinct in Azerbaijan; indeed, the historical memory of it is very much alive and relevant as communities negotiate new water distribution systems on rehabilitated *karezes*. Many well-known *karezes* continue to trickle along despite abandonment. A community member in Sinag Kahriz reported that, even in the military more than a decade ago,

> I would at times wake up on Saturday nights wondering if my brothers or somebody in my house back in the village was on time to open the water inlet on my fields, to claim our share of the kahriz water.

Some community members remember the zone of exclusion (*harim* rule) around the motherwell of a *karez* (no tubewells were allowed within a specified distance of the motherwell) and prohibition against extraction of water directly upstream from the exit point. This rule, designed to prevent groundwater-polluting activities, is particularly important in terms of maintaining water quality.

Because of abandonment during the Soviet era, many people have come to live around and over *karez* wells. Some people have *karez* maintenance wells in their backyards, and, in certain cases, they had started pumping water directly from those wells. Since the water rights on a *karez* are invariably adjudicated downstream from the exit point, extracting water directly from maintenance wells is tantamount to stealing water from all the shareholders in a *karez*. Communities often forcefully intervened to stop this practice, though in some instances they had allowed water users to extract water manually from some of the maintenance wells, and only for domestic water use purposes, provided the well was located on the person's property and there were no adverse effects on the quality and quantity of water at the daylight point.

A remarkable aspect of *karez* resurrection in the Nakhchivan region has been the organic reemergence of institutions one finds in the continuously existing *karez* system in Balochistan. For example, the position of *hisabgar*, which serves the same purpose as *mir-e-aab* in Balochistan, has been activated in Nakhchivan to adjudicate water rights and manage *karez* maintenance. On a different aspect, in a discussion during my first visit to Azerbaijan, I mentioned to one of the *karez* communities in the Ourdabad region that communities in Balochistan held two to three days of water right in reserve to be auctioned at the beginning of the cropping season to raise funds for *karez* maintenance. During my second visit a year later, I was pleasantly surprised to learn that the communities in Ordubad had instituted a water reserve, as had many other *karez* communities across the Nakhchivan region for purposes of auction or to supplement the supplies of water users facing a water deficit in any given season. Unlike in Balochistan, Ourdabad communities did not have active water markets, but there was a strong moral economy around *karezes*, whereby engaging in water sharing to supplement water-poor shareholders' supply was almost considered an obligation.

In all of the Nakhchivan region, the IOM project could only find one *karez kankan* (a person specializing in *karez* maintenance and locally pronounced chanchan), and he happened to be eighty years old. With the help of this elderly *kankan*, the project trained more than eighty *karez* workers, equipping them with modern safety gear to help approach the potentially dangerous task of *karez* maintenance safely.

One of the largest obstacles to *karez* rehabilitation that I observed in Nakhchivan was community discord. Yet I also observed progressive aspects of *karez* resurrection in the region, including the bestowal of water ownership to women, who at times even assumed leadership roles within their *karez* communities. While this progressive deviation from the historically patriarchal ownership pattern was partially a function of Soviet rule's gender-progressive legacy in the region, it also reflected the fact that most young men in the region had emigrated to Moscow, Baku, or even Istanbul to earn a living. Those left behind were primarily elderly men and women as well as young women and children. A comparable situation also existed in Balochistan, but not to the same extent as in Azerbaijan. It is in this context of prevalent rural-to-urban migration that the future of the *karez* system must be understood.

Local *karezes* provide these communities what the globalized labour market cannot—food security, identity, social capital, and cultural grounding. Given this context, *karezes*' key role and the motivation for engaging in their resurrection was fundamentally to resurrect the material basis for the existence of rural communities in Nakhchivan. The *karez* system's ability to provide a sense of certainty in an uncertain world—made somewhat more uncertain by climate change and neoliberalism—offers a true measure of *karez's* value to human societies. The sense of certainty provided by the *karez* is not limited to this particular traditional water system. Even in the most prosperous and globally hegemonic society, the United States, traditional *acequia* systems constitute a locus of community resilience. It is to that discussion we now turn.

Community before property: the subversive case of *acequias* in the southwestern United States

The first major waves of white Anglo settlers in the western United States were precipitated by the mid-nineteenth-century discovery of gold in California. The first white settlers in the Rocky Mountain West were primarily mining prospectors and homesteaders. In this arid region, mines and farmlands were not necessarily located near plentiful water sources and, hence, diverting water from its source became the norm. This was particularly true of water use by the economically dominant mining enterprises in the region. Given the arid geography, with water sources few and far between, the traditional riparian doctrine prevalent in the eastern United States and most of Europe, whereby only lands abutting water sources had a water right attached to them, was not very practical. With many water users seeking to divert from the same limited water source at any given time, questions of rights and priorities came to the fore and were resolved by the principle of first in time, first in right.

What started off as a geographically enforced innovation assumed the force of law by way of the Mining Act of 1866 and subsequently the 1870 amendment to the Mining Act, whereby prior appropriation (first in time, first in right) became the dominant legal doctrine in the American West. Prior appropriation was possible in the western United States because almost all the land in this part of the country was federally owned by virtue of the Louisiana Purchase, and every private person was considered a settler on it. The Desert Land Act (1877) ensured that water from all non-navigable water sources on public lands (almost all of the western United States) was available for appropriation for beneficial use (Getches, 1997).

The prior appropriation doctrine remains the dominant legal doctrine in the Rocky Mountain West today. The coastal states of California, Oregon, and Washington tend to follow a hybrid legal doctrine of prior appropriation and riparianism, along with fourteen other states that apply the prior appropriation doctrine to varying degrees. Prior appropriation dictates that in the case of water scarcity, the first person to divert water for beneficial use has priority of use over the person who may have appropriated water later in time. This doctrine divorces place of use from source of use. In its purest form, applied in most Rocky Mountain states, water rights can be bought and sold independently of land. The ability to buy and sell water is tempered by the no-injury rule, whereby other water users must not be adversely affected by the transaction.

The beneficial use concept has generally been narrowly defined under the prior appropriation doctrine, and instream water use, without diversion (especially by private non-state appropriators), has generally not been recognized as beneficial, with adverse consequences for Native American water rights and ecological quality (Getches, 1997). However, a greater accommodation of instream water uses for ecological purposes has emerged in contexts where water is appropriated and administered by the state, but not by private entities (see Merriman & Janicki, 2010).

Despite the ascendancy of the dominant prior appropriation doctrine as well as numerous cultural and economic changes over the past century and a half, the *acequia* communities of the southwestern United States have remained remarkably resilient. In fact, research by Ingram and Brown (1987) as well as by Rodriguez (2007) and Arnold (2008) documents that the key guiding principle of *acequia* communities is equity, where everybody carries an obligation towards maintenance of the irrigation infrastructure and in turn has a right to share in the bounty of water during wet years—as well as an equitable burden of the water scarcity during lean years. Members of *acequias* not only engage in a transactional relationship, but they share an identity and a sense of community that has crystalized around the water resource and its management (Arnold, 2008). The *mayordomo*, comparable to the *mir-e-aab* as discussed in the context of the *karez* communities in Balochistan, is the community leader who manages water distribution, adjudicating conflict and ensuring that everybody's rights and responsibilities are observed within the community.

Traditional agricultural and mining interests within the U.S. West have been the prime beneficiaries of the prior appropriation doctrine in the U.S. West. Though these interests have lost the clout they once enjoyed in the region's economy, they continue to hold disproportionate political power. Individualism and private enterprise as operationalized within the prior appropriation doctrine have become the locus of core identity politics among Anglo farmers in the American West. Water is the most litigated resource in this region (Bates et al., 1993). In fact, adversarial litigation is virtually inherent to the doctrine's structure. Beyond individual litigation, however, there is also inter-sectoral conflict over water, primarily between agricultural and mining interests on the one hand and the leisure, real estate, and Native American interests on the other. The prior appropriation doctrine and its underlying principles are therefore jealously defended by farming and ranching interests—they being the mostly senior appropriators. The same interests view environmentalist and communitarian agendas with considerable hostility. It is in this political context that the status of *acequia* communities in the U.S. West must be understood.

Acequias' ethos of shared scarcity and surplus runs counter to the individualistic ethos of the prior appropriation doctrine. Under prior appropriation, the senior appropriator can shut down junior appropriators to realize her/his share. Legally, the *acequias* were only recognized by the State of Colorado in 2009. Prior to then, only individual rights of the *acequia* shareholders were legally recognized, and their subscribing to the *acequia* norm of sharing scarcity instead of shutting down junior appropriators was purely voluntary (Hicks & Peña, 2012). Also legally problematic was the *acequia* norm of *one farmer, one vote* instead of a system in which voting rights are weighted proportionally to water right. Furthermore, while the prior appropriation system holds to a very narrow definition of beneficial use, typically involving direct agricultural or industrial productivity, documentation reveals that *acequia* communities explicitly recognize the more intangible value of water in terms of cultural and community coherence and ecosystem services.

An illustrative example of different values of water is the often repeated conflict between state and federal agricultural authorities and *acequia* communities regarding lining of the traditional earthen ditches for water conveyance. The state authorities have typically deemed seepage losses from unlined ditches in San Luis Valley in southern Colorado, for example, as a waste and have asked the communities to line them as a precondition of providing agricultural extension services or even enforcing their water right. The communities, however, have resisted those moves on the plea that, in fact, seepage from the channels provides valuable groundwater recharge, contributing to the formation of ecologically significant wetlands that are important to the overall health of the agro-ecological systems in which they operate (Hicks & Peña, 2003). With the ascendancy of more ecological thinking in the society at large, and even within state and federal government agencies, there is now greater sympathy to *acequia's* traditional irrigation technology.

The *acequia* communities of San Luis Valley have not only formed into networks of associations but have leveraged their moral economy to assimilate successive waves of new immigrants such that they maintain a "sense of attachment and mutual obligation by which they recognize themselves as members of a community" (Arnold, 2008, p. 41). In addition, the very same moral economy helps them negotiate to protect themselves against new institutions and sophisticated water appropriators, for example American Water Development Inc. (AWDI), a corporate entity that attempted to export groundwater from the San Luis Aquifer to the densely populated front range cities in Colorado (eastern slope of the Rockies). This attempted export was successfully opposed by a coalition of *acequia* and civil society actors through litigation and political mobilization over eight years between 1986 and 1994 (Nichols, Murphy, & Kenny, 2001). The judicial and legislative defeat of water transfers by AWDI was also a setback for the basic principle of prior appropriation, which does not limit the place of water use within state boundaries. The notion of the protection of local watersheds and regions and place-specific geographies of water is being rediscovered as a result of environmental and civil society activism (Wescoat, 1993).

In a changing West, the *acequia* communities uncannily encapsulate the post-industrial set of emergent ecological values gaining currency among certain constituencies. The articulation of those values may be in a somewhat limited biophysical environmental idiom, but even within that imperfect environmental idiom, a serious challenge to the established property interests in water and the established power structures in the West is going forth. Although the *acequia* ethos indeed centers on social justice and equity, remarkably, that very same ethos resonates with the dominant Anglo, biophysically-based environmental idiom. Emerging alliances between Anglo-environmentalists and Hispanic *acequia* communities will be important in future water politics of the U.S. West. The *acequia* communities provide a template showing how pioneers of western settlement such as John Wesley Powell envisioned cooperative action rather than competitive individualistic regimes as the basis for human settlement, not the competitive individualistic regimes in water management that actually came to pass. Therein lies the relevance of the *acequia* communities, like the *karez* communities, for the future of the regions where they exist.

Conclusion: familiar values, different futures

Connection, community, cooperation, and collective action are visceral human attributes. Thanks to these attributes, the human species has survived and thrived. But under modernity, these attributes have been replaced by competitiveness, profit, and accumulation as the dominant values underlying water management paradigms. It is little wonder, then, that contemporary water management regimes are socially divisive and, hence, ecologically unsustainable, as illustrated by the case of tubewells in *karez* systems and prior appropriation in *acequias*. Humanity faces many messy problems, not the least of which is climate change. But as some contemporary researchers have helpfully reminded us, to think of climate change as a future biophysical challenge is perhaps futile (see Hulme, 2009; Wescoat, 1991). The problem of climate change, like many others, must be recast as a moral and cultural problem, urgently relevant to here and now and not merely the future. The future stands on the foundations of the present, and the present state of water management is one of water appropriation by the powerful few at the expense of the many weaker segments of the society as well the non-human world. Climate change will not harm the weak and dispossessed of the world, and the non-human world, in ways that capitalism and predatory consumptions patterns of the powerful haven't done already. The perverse present has to be addressed not just by changing behavior, but by more fundamentally altering social practice. Marginalized water management techniques of the past and present such as the *karez* and *acequia* systems discussed in this chapter provide a clue to where we might begin.

Works cited

Arnold, T. C. (2008). San Luis Valley and the moral economy of water. In Whiteley, J. M., Ingram, H., & Perry, R. W. (Eds.), *Water, place and equity*. Cambridge, MA: MIT Press.

Bakker, K. (2003). Archipelagos and networks: urbanization and water privatization in the South. *The Geographical Journal* 169(4): 328–341.

Bates, S. F., Getches, D. H., MacDonnell, L. J., & Wilkinson, C. F. (1993). *Searching out the headwaters: Change and rediscovery in Western water policy*. Washington, DC: Island Press.

Beaumont, P., Bonin, M., & McLachlan, K. (1989). *Qanat, kariz and khettara*. Wisbech, UK: Menas Press.

Birkenholtz, T. (2009). "Irrigated Landscapes, Produced Scarcity, and Adaptive Social Institutions in Rajasthan, India." *Annals of the Association of American Geographers* 99(1): 118–137.

Brown, F. L. & H. Ingram (1987). *Water and poverty in the southwest*. Tucson, AZ: University of Arizona Press.

Budds, J. (2004). Power, nature and neoliberalism: The political ecology of water in Chile *Singapore Journal of Tropical Geography* 25: 322–342.

Getches, D. (1997). *Water law in a nutshell*. St. Paul, MN: West Publishing Co.

Hicks, G. A. & Peña, D. G. (2003). Community *acequias* in Colorado's Rio Culebra watershed: A customary commons in the domain of prior appropriation. *University of Colorado Law Review, 74, 2,* 387–486.

Hicks, G. A. & Peña, D. G. (2012). Conflicting norms in the struggle for communitary governance in the *acequias* of the Upper Rio Grande watershed. In Peña, D. G., Martinez, R., & Gallegos, J. (Eds.), *Voices of water, spirits of land: Four hundred years of acequia farming in the Upper Rio Grande Bioregion*. Tuscon, AZ: University of Arizona Press.

Hulme, M. (2009). *Why we disagree about climate change: Understanding controversy, inaction and opportunity*. Cambridge, MA: Cambridge University Press.

Kahlown, M. A. & Hamilton, J. R. (1994). Status and prospects of *karez* irrigation. *Water Resources Bulletin, 30, 1,* 125–134.

Lahiri-Dutt, K. (2003). People, power and rivers: Experiences from the Damodar River, India. *Water Nepal: Journal of Water Resources Development, 9, 10.5,* 251–267.

Lightfoot, D. R. (1996a). Moroccan khettara: Traditional irrigation and progressive desication. *Geoforum, 27, 2,* 261–273.

Lightfoot, D. R. (1996b). Syrian qanat Romani: History, ecology, abandonment. *Journal of Arid environments, 33,* 321–336.

Loftus, A. (2007). Working the socio-natural relations of the urban waterscape in South Africa Mar *International Journal of Urban and Regional Research*. 31(1): 41–59.

Merriman, D. & Janicki, A. M. (2010). Colorado's instream flow program—How it works and why it's good for Colorado. Retrieved from http://cwcb.state.co.us/environment/instream-flow-program/documents/whyisfprogramworksgoodforcopdf.pdf.

Mustafa, D. (2002). To each according to his power?: Participation, access and vulnerability in irrigation and flood management in Pakistan. *Environment and Planning D: Society and Space*, *20, 6,* 737–752.

Mustafa, D. (2013). *Water resources management in a vulnerable world: The hydro-hazardscapes of climate change*. London, UK: I.B. Tauris.

Mustafa, D. & Qazi. M. U. (2007). Transition from *karez* to tubewell irrigation: Development, modernization and social capital in Balochistan, Pakistan. *World Development*, *35, 10,* 1796–1813.

Mustafa, D. & Reeder, P. (2009). People is all that is left to privatize: Water supply privatization, vulnerability and social justice in Belize City, Belize. *International Journal of Urban and Regional Research*, *32, 3,* 789–808.

Nichols, P. D., Murphy, M. K., & Kenny, D. S. (2001). Water and growth in Colorado: A review of legal and policy issues. Natural Resources Law Center, Boulder, CO: University of Colorado School of Law.

Qureshi, A. S., Shah, T., & Akhter, M. (2003). The groundwater economy of Pakistan. Working Paper 64. Lahore, Pakistan: International Water Management Institute.

Rodriguez, S. (2007). *Acequia: Water sharing, sanctity and place.* Santa Fe, NM: School of American Research Press.

Sneddon, C. & Fox, C. (2006). Rethinking transboundary waters: A critical hydropolitics of the Mekong Basin. *Political Geography*, *25, 2,* 181–202.

Sultana, F. & Budds, J. (2013). Exploring political ecologies of water and development. *Environment and Planning D: Society and Space*, *30, 2,* 275–279.

Swyngedouw, E. (1997). Power, nature and the city: The conquest of water and the political ecology of urbanization in Guayaquil, Ecuador: 1880–1990. *Environment and Planning A*, *29, 2,* 311–332.

Swyngedouw, E. (1999). Modernity and hybridity: Nature, regeneracionismo, and the production of the Spanish waterscape, 1890–1930. *Annals of the Association of American Geographers*, *89,* 443–465.

van Steenbergen, F. & Oliemans, W. (2002). A review of policies in groundwater management in Pakistan 1950–2000. *Water Policy*, *4, 4,* 323–344.

Water and Power Development Authority (1992). Uncontrolled mining of groundwater in Balochistan and possible remedial measures. Quetta, Pakistan: WAPDA Hydrogeology Project.

Wescoat, J. L. Jr. (1991). Managing the Indus River basin in the light of climate change: Four conceptual approaches. *Global Environmental Change: Human and Policy Dimensions*, *1, 5,* 381–395.

Wescoat, J. L. (1993). Water law, urbanization and urbanism in the American West: The "place of use" reconsidered. *Urban Geography*, *14, 5,* 414-–20.

Zeitoun, M. & Warner, J. (2006). Hydro-hegemony, A framework for analysis of transboundary water conflict. *Water Policy*, *8, 5,* 435–460.

Zeitoun, M. (2010). Environmental geopolitics and hydro-hegemony: the case of Palestine-Israel In : History of Water Series II, Volume 3: Water, Geopolitics and the New World Order. I.B.Tauris & Co Ltd.

Zeitoun, M., Mirumachi, N., & Warner, J. (2011). Transboundary water interaction II: The influence of "soft" power. *International Environmental Agreements: Politics, Law and Economics*, *11, 2,* 159–178.

12 The politics of muddled waters in Gujarat

A religious nationalist development model's treatment of water

Pushpa Iyer

Water in Gandhi's land: an introduction

Gujarat is the land of Mahatma Gandhi, who formed his ashram on the banks of the River Sabarmati in Ahmedabad city and from where he proclaimed and lived his message of *satyagraha* (the struggle for truth) and *ahimsa* (nonviolence).[1] Gandhi's ashram was a sacred place where he spent hours with his followers in prayer and meditation, in deliberations, and in discernment as the calm waters of the River Sabarmati flowed past. Listening to the flow of the River Sabarmati, Gandhi spoke these famous words: "The earth, the air, the land and the water are not an inheritance from our forefathers but on loan from our children. So we have to handover to them at least as it was handed over to us" (Kaushik, 2010).

Today, the banks of the River Sabarmati are vastly different from the banks on which Gandhi led prayers. Many decades later, water released from a mega-dam flows into the river, the banks are concrete, and the poor no longer have access to the riverbank which once sustained their livelihoods. Instead, the river serves the purposes of middle-class entertainment. Customs and traditions in the state include offering water to the Sun God every morning, and water gathering places are central in every community. Rural communities feature watering holes for animals, yet communities deny each other the commodity of water simply because of identity politics. So, what is the water story in Gandhi's land? How well have its citizens handled this resource, and in what condition will they hand it over to the next generation?

This chapter argues that the religious-nationalist party, the Bharatiya Janata Party (BJP), has furthered caste, class, and gender divisions and that its development paradigm, under which its water policies have been drawn, is designed to perpetuate these divisions. Arguing that most water scarcity in Gujarat has been political, this chapter details the water situation in contemporary Gujarat, focusing on the period between 2002, a year of large-scale state-sponsored religious violence, and 2014, when the state's chief minister became the country's prime minister. Governmental water policies have discriminated against the poor and the marginalized while benefitting the rich and the privileged. In short, the societal structures that marginalize and disfavor some while preferring others are the very same structures that control resources and hold the power to make decisions about water supply, accessibility, and distribution. Gandhi's desire that water, a basic human need, be treated as a gift loaned to us for safekeeping on behalf of future generations but today has ended up in the hands of political parties and used only as a tool to reach political objectives.

Gujarat: a water-stressed state in India

With a population about 60 million, Gujarat, in northwestern India, is the tenth most populated state in the country (see Figure 12.1). The state spans an area of about 190,000 square kilometers, making it the seventh largest in the country (Government of Gujarat, 2011). However, while Gujarat boasts 6.39 percent of the country's geographical area, it contains a mere 2.28 percent of India's water resources (Gupta, 2011). The main source of water for Gujarat is surface water. The state controls 185 river basins, of which there are only eight perennial rivers, all of which are located in the south. Of all the rivers in Gujarat, 168 are in drought-prone parts, and each of these is non-perennial (Barot, 1996; Gupta, 2011). Unequal distribution of water resources to the different parts of Gujarat extends to the rainfall in the state. While southern regions receive a rainfall of about 1107 mm, the Saurashtra (western) and Kutch (northern) regions receive only 578 mm of rain annually[2] (Upadhyay, 2004).

Gujarat records indicate that in a ten-year cycle, every three years have been drought years. Since India's independence in 1947, Gujarat has seen over twenty years of officially declared drought (Gupta, 2011).[3] As the people of Gujarat often say, having a drought-free year is much rarer for the people of this state. In fact, the recently published UN World Water Development Report of 2014 declared Gujarat one of the most water-starved states in India. It would seem that Gujarat is fighting the inevitable. As long as it remains a drought-prone state, it must spend much money to unearth water resources, exploiting nature in the process.

Gujarat's history, however, tells a different story. Dholivara, an archaeological site in the Kutch district of Gujarat, contains the ruins of an ancient Indus Valley civilization, Harappan city. One of the unique features of Dholavira is its sophisticated water conservation system of stone channels and reservoirs, the earliest found anywhere in the world. The city featured massive reservoirs, three of which are exposed. They were used for storing fresh rain water or water diverted from two nearby rivulets, Mandsar and Manhar. One gets the idea that this ancient city made efforts to conserve every drop of water. House drains emptied into main drains, which ran under the streets. Drainage systems were covered with stones and wood with inspection traps. This sophisticated system bears testimony to its users' heightened knowledge and awareness of sanitation and health (Kulkarni, 2013).

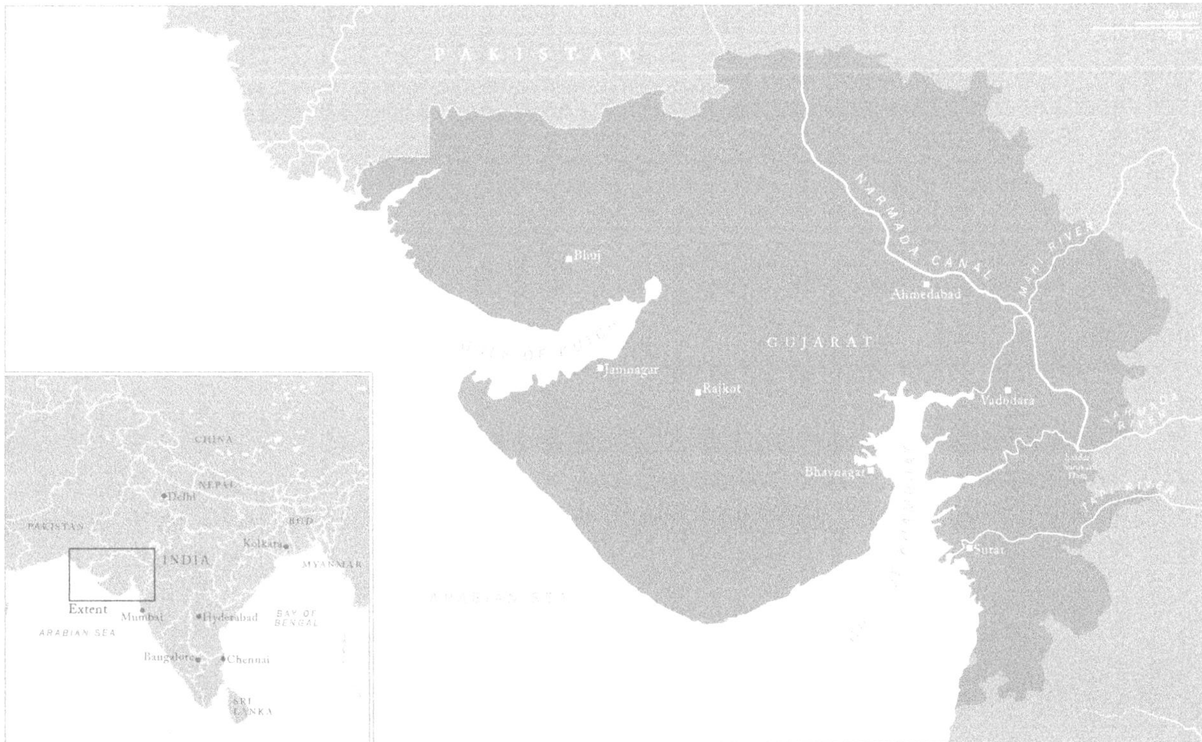

Figure 12.1 Map of Gujarat showing the stretch of the Narmada Canal across the state. The north and northwestern parts of the state are the regions of Saurashtra and Kutch, the most drought prone regions

Source: Cartography by Gregory Woolston using data from Investment Innovation & Research Alliance, retrieved from http://www.iira.in and Natural Earth.

"The kind of efficient system of Harappans of Dholavira, developed for conservation, harvesting and storage of water speaks eloquently about their advanced hydraulic engineering, given the state of technology in the third millennium BCE," says R. S. Bist, Joint Director General (Rtd.), Archeological Survey of India. In modern times, Gujarat seems to have lost its ability to develop and maintain such intricate yet socially and economically viable systems despite the fact that the climate, topology, ecology, and precipitation in the region of Kutch have changed little since the days of the Indus Valley civilization.

While it would be true to say that Gujarat suffers water scarcity, this could be read as an oversimplification at best and completely irresponsible at worst. Why? Because water scarcity should always be considered alongside issues of unequal access, distribution, and supply. The situation in the state is more akin to what Schultze (2014) at the Pacific Institute defines as a "water stress." According to him,

> "Water stress" refers to the ability, or lack thereof, to meet human and ecological demand for water. Compared to scarcity, "water stress" is a more inclusive and broader concept. It considers several physical aspects related to water resources, including water scarcity, but also water quality, environmental flows, and the accessibility of water.
>
> (Schulte, 2014)

Inequality in Gujarat is complex, originating in high levels of historical, social, and economic disparity, which in turn are bound to power and manipulation. In particular, stratification by caste, class, and gender, which is endemic to India, is prominent in Gujarat, and each one of these stratifications has a strong relationship with water (Upadhyay, 2004). Religious divisions, which closely connect with the ruling party's political agenda, constitute an additional divide.[4] The urban–rural divide, the north–south divide, and others are examined below.

Water in the context of urban–rural, caste, class, and gender divisions

According to the 2011 census, 42.6 percent of the state's population is urban, and 57.4 percent of its population is rural (Government of Gujarat, 2011). Since the previous census in 2001, around 5.2 percent of the population appears to have moved from rural to urban areas (Government Gujarat, 2001). While numerous reasons factor into this migration, water or the lack of it played an important role in many people's decision to move.

Rural populations and their water needs, therefore, deserve more attention if for no other reason than the fact that their needs vary considerably from the needs of the urban population. Of the 6188 villages in Gujarat, about 34.3 percent are no-source villages, that is, they have no definite source of water (Gupta, 2004). Most of these villages are in the Kutch and Saurashtra regions in the northwestern part of the state.

The rural population (60 percent of the state's total population) is overwhelmingly (86 percent) involved in the agriculture sector. Of these, over 60 percent are Dalits—the untouchable caste group of India[5] (Government of Gujarat, 2011). Further, the census data shows that there are 54.47 lakh[6] "cultivators," or farmers, in Gujarat, and an additional 68 lakh agricultural laborers (Nair, 2013). Groundwater constitutes the major source for both irrigation and domestic use in most parts of rural Gujarat. In fact, 80 percent of irrigation in the state is via groundwater. There has been an increase in irrigation wells and tubewells in different parts of the state, which also illustrates the state's dependence on groundwater for irrigation (Gupta, 2004).

Now, if one considers the fact that an overwhelming majority of the state's agricultural laborers are landless Dalits and that the few Dalits who might be cultivators are marginal landowners (Soni, 2006), it is not hard to infer where the control of water lies. Water sources (wells, tubewells, and borewells) are all usually located on the farms of the upper castes, making Dalits, the landless laborers, completely dependent on the higher-caste groups for their basic needs, indeed for their survival. Upper-caste communities or families decide how much water they'd like to share with a Dalit community. Dalit women, who walk long distances to gather water for their families, are often harassed on these treks, and become victims of various kinds of violence, including sexual violence.

Typically, the inhabitants of slums and other poor areas in cities do not have access to water sources but rely completely on local authorities for their water supply. Once again, a high percentage of the Dalit community lives in the urban slums, and once again, women and girls carry the primary responsibility for gathering water. Apart from harassment, women and girls are often forced to give up social and economic pursuits in order to ensure water resources for their family. Further, in both rural and urban areas, development planners assume that women's only interest in water is for its domestic uses, while men are concerned with its productive uses.[7] This assumption not only disadvantages and underestimates women, but it also leads to unsustainable development projects (Upadhyay, 2004).

Water suppliers, that is, the private tanker companies that bring in water, well recognize the costs of water. These private companies usually buy water from higher-caste members and the more economically advantaged who have an abundance of it. The companies then sell the water to the poor and marginalized at exorbitant prices. Local authorities' failure to ensure water for all gives power to these private companies. Whether in the hands of private companies, landowners, local authorities, or the upper castes and the rich, the drive to control and manage water sources becomes necessary to retain the power that their economic, social, and political status brings. But power over one of the most basic human needs is almost always an abusive and manipulative power, making violence against the less powerful systemic (Galtung, 1969).

Systemic violence then becomes part of the structure and serves to maintain it, which in turn continues to deny opportunities to the disadvantaged, keeping them in a status quo limbo or pushing them down further. In a case study of village Vadali (Chotila Taluka, Surendranagar district), Prakash and Sama (2006) detail the social economic conditions of the impoverished and landless Dalit farmers and their struggle for water vis-à-vis the higher-caste community, the Ahirs, who own land and have wells on their farms. When drinking water sources dried in the village, the Ahirs refused to permit the digging of a new drinking water well in the village simply so they would not have to share the aquifers with additional wells. As a result, the Dalits, who are socially barred from drawing water from the wells of the Ahirs, were forced to walk long distances to gather water. The Dalits never protested against this injustice, as the rigid caste system prevents the Dalit community from openly voicing their protest against members of a higher caste. Additionally, the Dalits tend to accept their situation as their fate and not fight it. However, a local nongovernmental organization (NGO) working in the region noticed the plight of the Dalits and turned to the local government for help. To find a solution to this problem, the Water and Sanitation Management Organization (WASMO) decided to find a different location for the drinking water well, one that would require de-silting another check dam (Prakash & Sama, 2006). In the end, the system enabled the Ahirs to maintain control over their water and ensured that every decision about water needs and access pivoted on their power. Oppressive structures thus perpetuated the status quo, preserving the power of the Ahirs over the Dalits.

In short, economic, social, and cultural divisions play a role in access to and in the distribution of water in Gujarat. The resultant control over water sources makes water a political and economic commodity for the powerful and a social, cultural, and basic need for the underprivileged.

From Gandhi's land to Modi's land: religion as another division in Gujarat[8]

Gujarat's population of 60 million represents a variety of ethnic and religious identities. Hindus comprise 88 percent of the population, Muslims 9 percent, and 3 percent of the population is a mix of Christians, Jains, Sikhs, and Parsis. Tribals, who comprise 14 percent of the population, are spread over the eastern borders and share cultures across borders, and they do not adhere to a single Gujarati tribal identity. The regions of Kutch and Saurashtra boast their own unique culture and language, having existed as independent units before Gujarat became a state in 1960. Diversity in Gujarat extends to languages (dialects) spoken, food (various styles), dress (depending on one's caste or ethnic identity), and geography (desert in the north and forest in the south).

Ironically, Gujarat, where Gandhi preached "unity in diversity" and nonviolence, is also infamous for regular bouts of violence—riots—between the Hindu and Muslim communities, each with its own distinct culture, socio-economic opportunities/conditions, language, and history. The conflict lies in this diversity.

The conflict between Hindus and Muslims in this region deepened when the BJP came to power in the mid-1990s, with their religious-nationalist ideology, further politicizing the conflict between the two religious communities.

Supported by other socio-religious groups collectively known as the Sangh Parivar, the BJP advocates a fascist ideology—Hindutva—*One nation, one language, one culture, one people*, effectively discriminating against and oppressing Muslims and other minorities. Furthermore, the version of Hinduism under Hindutva is Brahmanical (upper caste), which has no place for the Hindu traditions of the lower castes (such as the Dalits).

Since coming into power, the BJP, led by Narendra Modi, has used Gujarat as a Hindutva laboratory for their experiment of making a "pure" land. The culmination of their experiment took place between February 28 and March 2 of 2002, during which over 2000 Muslims were killed, and hundreds of thousands were dispossessed and displaced by "Hindu" mobs. The violence was framed as Hindu mob violence in retaliation to a Muslim mob's burning of a train compartment (carrying Hindu religious workers). However, the scale of Hindu participation was not only disproportionate to Muslim participation, but it was so one-sided that the riot did not resemble any of the previous riots in Gujarat between the two communities. Evidence of state complicity in the violence became more and more obvious with every investigation, prompting many to call the events a pogrom, ethnic cleansing, or even genocide.[9]

Modi, the architect of the 2002 violence, has never shown any remorse, remaining dispassionate, uncompassionate, and strategic in denying justice to victims. All accusations against him have been swiftly dismissed through force. As Modi revamped his strategy to make Gujarat and himself a symbol of development, centering security, development, and even Hinduism on himself, there is no doubt that the events of 2002 transformed Gujarat from Gandhi's Land to Modi's Land (Iyer, 2013).

The social divisions in Gujarat and their political manipulations created a pre- and post-Modi Gujarat. While these divisions existed previously, they had not defined the state and all the policies until after the BJP came to power. Even more striking is the state's attempt to provide solutions for all issues using a social system that disfavors anyone who is not a Hindu, an upper-caste Brahmin, and economically well-off.

So where did this leave Muslims in Gujarat, the state's second- if not the third-class citizens? Since the violence of 2002, they have come to inhabit ghettoized areas in cities. Most of the Muslim community became internally displaced following the violence, living in semi-permanent structures where government facilities are nonexistent. Water is their most important need, and one for which practically no infrastructure exists. While, for example, Ahmedabad city has seen massive infrastructure growth and

> has emerged as one of the most "preferred cities to live in" with uninterrupted power supply, sprawling shopping malls, multiplexes, flyovers, bus rapid transit system (BRTS), and high rise apartments, the city's infrastructure boom has hardly touched the Muslim community, who are forced to live in slum-like conditions—where there is no clean water, no sewage system or street roads.
>
> (Langa & Gadhavi, 2014)

A visit to some of these areas highlights a lack of drainage facilities. Ironically, one wades through muddied waters to reach homes that have no regular water supply. There is a sense of desperation in the people, who recognize that their identity (as Muslims) not only denies them access to water, a basic human need, but prompts many, especially young men, to discard their identity and assimilate into mainstream society. The decision to retain one's "oppressed" identity as a form of protest or to modify and reinvent that identity in order to survive is a difficult one that results in many internal conflicts in the community.[10]

Juhapura—a residential area of Ahmedabad city (also known as mini-Pakistan, labelled to signal "enemy" territory) —offers no piped water, and yet residents receive a water tax bill. Many rely on borewells, which only produce saline water (Peer, 2014). A recent survey of sixty-three Muslim-dominated areas in Gujarat revealed a high level of internal conflict over lack of water. Of the sixty-three locations, only sixteen received water from the municipality, and twenty-six received water from the village *panchayat*. Others depend on wells, borewells (mostly private), and, in big cities like Ahmedabad, erratic water supply from the Gujarat Water Board. In some areas of the Bharuch district, people have to walk more than one kilometer to obtain water (Counterview, 2014).

Structural discrimination over water: the Modi Model

The Modi Model of Development (MMD) defines post-2002 Gujarat. Modi and the BJP have successfully transformed the image of Gujarat from the land of genocide to the land of development. Key economists describe Modi as a marketing guru, not a development guru, one who successfully filled the void for justice with development fluff—a very visible, pleasing-to-the-eye kind of development emphasizing roads, bridges, buildings, and other infrastructure. They argue that the MMD is a classic neo-liberal development model benefitting the middle class and therefore widening the gap between rich and poor (see for example Hashmi, 2014; Sachitanand, 2013; Subramanian, 2013). The need for water and the right to water are cornerstones of any development policy. The following paragraphs address key aspects of MMD's treatment of water—a critical natural resource.

Water as a casualty of development

In a state heavily reliant on agriculture, the MMD de-stresses agriculture, especially small-scale agricultural practices. Despite constant propaganda advertising rapid growth in agriculture due to improved technological practices and crops such as Bt (Bacillus thuringiensis) cotton (IANS, 2014), the facts speak differently. The 2011 census results reveal that in the decade between 2001 and 2011, there was a decrease in the number of cultivators by 3.55 lakh and a simultaneous increase in the number of agricultural laborers by 16.78 lakh (Government of Gujarat, 2011). The rise in landless agricultural laborers suggests growing income disparity among the rural population. Land ownership is bound with the ability to own or buy water sources for the purposes of irrigation. Small landowners, as we all know, often suffer dispossession of their land and end up at the mercy of large landowners for whom they work, struggling to meet their basic needs, including water needs.

Reliance on groundwater for irrigation and the continued depletion of aquifers has led to increased investment in heavier capital-intensive submersible pumps and greater electricity demands in order to dig deeper—but with very little success. Kishore (2014) finds that poverty is significantly higher in such places and conflicts more frequent. In fact, he suggests a direct correlation between falling groundwater levels and increased levels of poverty. He adds that government subsidies and other programs have only benefitted the well-to-do farmers, not the poor and landless (lower-caste) farmers. Other scholarly research supports his findings (Kishore, 2014). For example, in Gujarat, farmers pay a flat rate for electricity irrespective of how much they use, and small farmers have no control over heavy equipment and extensive pumping for water (Narula, Fishman, Modi, & Polycarpou, 2011). Once again, when the imbalanced system favors the wealthy, emergent policies and practices further serve to disadvantage the poor.

Overreliance on groundwater leads to another problem: water salinity. Indira Hirway argues that northern Gujarat, and Kutch in particular, features favorable geomorphic conditions capable of retaining groundwater but agricultural practices have contributed to its severe depletion, and water tables have fallen because of the many industries that have sprung up along the coastline (Hirway, 2005). Gujarat boasts the longest coastline in the country (1600 kilometers), and the MMD's "port-led development" approach has seen massive growth of industries along the coast in detriment to traditional fishing industries, marine ecology, and water salinity. Effluents from the growing number of industries also contribute to increased pollution of the many rivers in the state (Dasgupta, 2010). This pollution is not surprising in light of the fact that these large coastal industries have taken control of existing water resources and have dug deeper and deeper into the ground to secure water.

According to one report, the decline of water tables over the years to about 195 meters below ground level has meant risking irreversible salinization of aquifers (Narula et al., 2011). In coastal areas, salinity has reached about 2000 parts per million (ppm) of totally dissolved solids (TDS) leading to an increase in "ghost villages" along the coastal belt because of large-scale migration to nearby towns and cities (Barot, 1996). Migration has often meant moving livestock, which in turn leads to shifts in the primary workforce and to economic, social, and cultural dislocation (Gupta, 2011). What is more, excessive fluoride and nitrates in the water, together with increased salinity, has caused health issues in many parts of the state.

The MMD's port-led development program is also insidious in its focus on the 1125-kilometer coastline of two of the most drought-prone and water-scarce regions of the state, Kutch and Saurashtra. Kutch views this development as a way of mainstreaming the region in an effort to become more "Gujarati"—part of the broader "One nation, one language, one culture, one people" program.[11] Ironically, as fresh water has become increasingly privatized, the poor and marginalized people of Kutch are finding themselves with diminishing access to water. In interviews, many fishermen of the Mandvi area recounted desperate searches for water. They often found themselves depending on the very companies responsible for destroying their livelihood through development-related destruction of ocean life. To cope, these fisherman must develop split personalities of a sort: they must appease the companies in order to receive water for survival, while at the same time joining forces with human rights organizations to stand up against corporate-led development along the coastline. This phenomenon makes social development interventions very complicated.[12]

In addition to water control and distribution issues, much water scarcity is due to the lack of importance given to the revival and modernization of traditional water systems for domestic use. The two western states of Gujarat and Rajasthan are well known for their stepwells (*vavs*)—so called because of the steps leading down to them. Stepwells emerged around the fourth century, providing year-round water to the community for drinking, washing, and irrigation purposes. Compared to other wells, stepwells were easy to access (because of the steps) and also easy to maintain. By the eleventh century AD, these landmarks widely dotted the region. However, when the British colonized India in the nineteenth century, they wanted to close down these wells, which they considered unhygienic. Thus began the decline of the stepwell, which had become an important place for social and cultural gatherings. Constructed under both Hindu and Muslim rulers, stepwells are engineering masterpieces and architectural wonders, with intricate carvings featuring the religious and cultural symbols of the time. Today, most stepwells are dry due to groundwater tapping (Bliss, 2014).

Village water tanks and ponds (*talavs*) have traditionally constituted additional sources of water, but, like stepwells, these too have decayed over the years due to lack of maintenance and overreliance on groundwater. As Das explains, ponds and tanks, like stepwells, could become sustainable sources of water if revived and maintained under local ownership. Unless ownership is decentralized and local, water sources will not be protected and managed (K. Das, 2009).

The MMD has ignored traditional water systems for fear of private companies who might lose control of water sources. Specifically, they fear yielding control of water—a precious economic commodity—to communities. Government programs and policies have done nothing to maintain traditional water systems—architectural and engineering masterpieces that served communities reliably for centuries.

In its search for external solutions to the water crisis, the MMD has preferred to spend money and energy developing major projects rather than investing in micro-level solutions at the local level (Mehta, 1995–1996). For a development model that prides itself on pretentious, eye-appeasing projects, this is not a surprising move, and it is no wonder that the Gujarat government manipulates perceptions of water scarcity to justify its massive investment in the controversial

Sardar Sarovar Project (SSP) (also known as the Narmada Dam project). The website for the project (http://www.sardarsarovardam.org/benefits-of-project.aspx) details scarcity, especially in the Kutch region, and frames the project as directly addressing this scarcity. The 163-meter dam is expected to provide water to over 30 million people and irrigate over 17.8 hectares of land in the state (Mehta, 1995–1996). Controversies over the dam's height, ecological impact, and its massive displacement of people, especially tribals, has delayed the project several times. Water is expected to enter the northern parts of the state only once the main canal has been constructed, traversing over 500 kilometers before it enters Kutch. Because the water will cross major rivers and the desert region of Kutch, much of it is expected to evaporate.

Built at enormous social, financial, and environmental costs, and primarily to meet the needs of central and urban Gujarat—specifically, the builders' lobby—the project rests on a platform of perceived dire water scarcity with a promise of meeting all of the state's water needs. Further, the MMD has manufactured consent from many sectors, pitching the SSP as the only solution for water scarcity in the state (Mehta, 2001). Meanwhile, people continue to queue in long lines and pay exorbitant amounts to private tanker companies for drinking water in drought-hit areas of Gujarat while the government of Gujarat sanctioned 3 million liters of water from the SSP to a Coca-Cola plant (Pinto, 2014).

Water for some only

For a state with acute water scarcity, Gujarat has devoted a surprising amount of effort to create infrastructure and opportunities for enjoying water. The state boasts numerous water parks for amusement—for those who can pay to enjoy them. Das reports that during the acute drought of 2000, the Gujarat government permitted these parks to remain open, succumbing to pressure from the powerful park owners (R. Das, 2000). For a hefty price, a small percentage of people continued to enjoy water, while a large number of people were left starving and had to focus their energies entirely on efforts to obtain drinking water, feed their cattle, and water their lands. This story repeats year after year as the state suffers from successive droughts. A government that turns a blind eye to its starving and thirsty people—the poor and marginalized, of course—is succeeding in perpetuating imbalanced societal structures.

The Sabarmati Riverfront program is yet another example of the MMD's deliberate insensitivity and forceful approach. Originally proposed by the Congress Party in the 1960s, the Modi government in Gujarat has overseen its building and completion over the past three decades. In the name of beautifying the city of Ahmedabad, the BJP has forcibly evacuated hundreds of thousands of families on the bank of the river Sabarmati, people who, for years, had depended on the river to meet their basic needs and support livelihoods such as small-scale gardening, local laundries, and regular markets. Riverbank residents also used the water as a venue for cultural activities, which have since been uprooted, along with the residents, and sent off to the city outskirts (Mathur, 2012).

The banks have been poured with concrete, high walls have been built, trees planted, waters released from the Narmada main canal, and pleasure boating facilities introduced, providing those with time to enjoy life a place to do so. Meanwhile, the poor and displaced have been reduced to commuting long distances into the city to look for casual laboring work. Mathur opines, "The riverfront project as it unfolds today is itself a totalitarian modernist planning project treating (riverbank) space as devoid of the cultural, social, economic and political elements, through which the urban working poor negotiates its place in the city" (Mathur, 2012).

Another major BJP government project has been the renovation and privatization of Kankaria Lake in Ahmedabad city. One of the largest artificial lakes in the city, Kankaria Lake was built in the fifteenth century by a Mughal ruler for bathing purposes. The lakefront project involved renovating the entire area into a pleasure grounds with water parks, amusement rides, street food, and more. The entry fee, of course, has made the lake inaccessible to the poor.

Skewed access to water for the poor and the marginalized is further evidenced by inequitable access for domestic purposes. In Bhuj, domestic use of water was 14 liters per capita per day (lpcd) in the slums, while it was 79 lpcd in middle-class housing societies and 108 lpcd in upper-class residential communities. In Ahmedabad city, water consumption was 5 lpcd in some slums and 500 lcpd in upper-class residential communities, for example in the Shahibaug area of the city (Hirway, 2005). The MMD has clearly left many out.

Structural violence, cultural violence, and water stress

The above examples make it clear that solutions to water stress in Gujarat are to be found in the very structures responsible for furthering social divisions. These structures are imbalanced to discriminate against, oppress, and deny opportunities to some, while furthering the benefits of others. Galtung (1969) argues that these imbalanced structures are inherently violent. Those against whom the structures discriminate, he argues, experience violence akin to direct physical violence, a phenomenon he terms structural violence. Naturally, then, structural violence will ensure that solutions to water scarcity, supply, distribution, and access originate within the structure and not from without.

The case of Gujarat is not only a case of structural violence against the marginalized—the Muslims, Dalits, rural populations, and the economically poor—but of culturally-sanctioned violence. Years of oppression have disempowered these communities to the point where they rarely raise overt resistance to individual cases of discrimination. Instead, they accept destiny. On the other hand, the majority communities, that is, the powerful and rich, continue to perpetuate structural violence, and the culture sanctions their role as perpetrators. A culture that frames all Muslims, Dalits, women, and the poor, as second- or third-class citizens who can be deprived of even water—a basic need—is a culture that supports and sanctions the ongoing violence, what Galtung calls cultural violence (1999).

Cultural violence and structural violence go hand in hand. In effect, both maintain the status quo, and it is in this context of social, economic, and political divisions that water policy is drawn and implemented. In such a context, can one expect the policy ever to be just? Structural and cultural violence is, in essence, the most formidable barrier to ensuring water for all in Gujarat.

When considering water as a human right, it is essential to understand the role of these two types of violence. The 2010 United Nations (UN) resolution calls upon "states and international organizations to provide financial resources, help capacity-building and technology transfer to help countries, in particular developing countries, to provide safe, clean, accessible and affordable drinking water and sanitation for all" (United Nations, 2010). By this standard, the government of Gujarat has clearly failed its people on many levels.

Conclusion

In Gujarat, custom holds that when a guest enters a home, the first thing the host offers is water. The state's topography and the climate demand this particular expression of hospitality, but it has become a unique symbol of Gujarati culture. This custom is special in that it does not discriminate against Dalits, Muslims, or any other group. Everyone is offered water, although discrimination may surface in the way the water is offered; for example, some households maintain separate cups for Dalits or pour water into the hands of the Dalits instead of offering them a cup of water in order to prevent them from "polluting" the host's utensils. How tragic that the politics of a nationalist-religious government has failed to build on this unique tradition of the state, but rather has led to denying entire villages, slums, and certain groups of people access to water.

Since 2002, Gujarat has become a far more divided state, and its political, social, and cultural divisions define water access. The MMD has directly impacted water control and water distribution. Religion, caste, and economics have all polluted the water, making it very difficult for many in Gujarat to survive (literally). The BJP government in Gujarat has not only disrespected our inheritance passed down from our forefathers in favor of profits, but it has decidedly misused our loan from future generations. Gandhi would be dismayed.

Notes

1 Ahmedabad city is the commercial capital of Gujarat State, located in the northwestern part of the state.

2 In Gujarat, often referred to as a water-stressed state (see for example Das, 2005), disparities are compounded by the unsurprising fact that groundwater depletion has far exceeded groundwater replenishment in all areas, but in particular, in the water-starved northern areas of the state. Here, over 3000 mcm of groundwater is annually pumped out against a replenishment of 2400 mcm, leaving a deficit of 600 mcm, which, not surprisingly, costs the economy Rs. 300 crore (Bhaduri, 2012). Further, overreliance on groundwater increases the water's salinity, especially in coastal areas.

3 According to Gupta's article, roughly twenty drought years occurred between 1947 and 2003, but other reports indicate a higher number. Since 2003, many more drought-hit years have passed, including 2013, which was one of the more severe ones.
4 As in the rest of India, there are many religious communities in Gujarat, and relationships between these religious communities is often fraught with divisions directly connected to political manipulations.
5 A very marginalized group, the Dalits have been discriminated against for centuries—socially, politically, culturally, and economically. Gujarat is no different in its treatment of Dalits. Prevented from being a part of mainstream society, Dalits are invisible to the privileged; they are treated as lesser beings.
6 1 lakh unit = 1,000,000 units.
7 Quantifications of water productivity (value of crops etc.) tend to disadvantage women, whose use of water is considered to be entirely for domestic purposes. Not including these domestic uses in water quantification efforts further marginalizes women. Water is an economic good and since women are already left out of water management decisions it is important that their primary use of water (for domestic purposes) is included when calculating water productivity who are at a great disadvantage when it comes to access, since neither technology nor human power are leveraged to develop just supply and distribution systems.
8 This section is mostly reproduced from the book *Gandhi's Gujarat: Vibrant diversity* by Pushpa Iyer.
9 The goal of the 2002 Hindutva experiment was to saffronize the state, saffron being the color of the Hindu nationalist parties and representing their fascist ideology—and the experiment was shockingly successful. The saffron hue spread to every level of state institution, including the "independent" judiciary, influencing both the concept of justice and the campaign for justice.
10 Based on interviews conducted by the author with Internally Displaced Persons (IDPs) of Juhapura in January 2013.
11 An important dimension of this development is the recent religious divisions in Kutch. Kutch has historically seen itself as separate, with its unique culture, food, language, clothes, and traditions. Using the guide of development, the government, in a bid to gain more inroads into Kutch, has succeeded in introducing Hindu nationalism in the region. Kutch has in recent years seen many Hindu–Muslim clashes—a completely new and worrying experience for the people of Kutch.
12 Interview with fisherman in Mandvi in Kutch, 2012.

Works cited

Barot, J. M. (1996). Occurence of salinity in Gujarat. Paper presented at Reaching the Unreached: Challenges for the 21st Century, New Delhi, India.

Bhaduri, A. (2012). Rivers of Gujarat: Many variations and contradictions: A lecture by Himanshu Thakkar at the Living Rivers, Dying Rivers Series. India Water Portal.

Bliss, L. (November 21, 2014). Can India's ancient stepwells help solve the country's water crisis? CityLab (The Atlantic). Retrieved from http://www.citylab.com/design/2014/11/can-indias-ancient-stepwells-help-solve-the-countrys-water-crisis/383045/

Counterview. (January 13, 2014). Most Muslim areas of Gujarat lack basic infrastructure facilities. Retrieved from https://counterview.org/2014/01/13/most-muslim-areas-of-gujarat-lack-basic-infrastructure-facilities-whether-water-sanitation-health-or-education/

Das, K. (2005). Drinking water and sanitation in Gujarat: Crisis and response. In L. Visaria (Ed.), *Gujarat's development by the mid 2000s: A compendium*. Ahmedabad, Gujarat: Gujarat Institute of Development Research.

Das, K. (2009). Village water tanks in Gujarat's desert region: Potential and relevance. *International Journal of Rural Management, 5, 2*, 129–144.

Das, R. (April 27, 2000). Water parks drought-proof, courtesy Govt. of Gujarat. *The Hindustan Times*.

Dasgupta, M. (June 14, 2010). Industrial effluents polluting Gujarat rivers, says forum. *The Hindu*.

Galtung, J. (1969). Violence, peace, and peace research. *Journal of Peace Research, 6, 3*, 167–191.

Galtung, J. (1999). Cultural violence. In M. B. Steger & N. S. Lind (Eds.), *Violence and its alternatives: An Interdisciplinary reader.* New York: Palgrave MacMillan.

Government Gujarat. (2001). Gujarat census report. Retrieved from http://censusgujarat.gov.in/

Government of Gujarat. (2011). Gujarat census. Retrieved from http://www.census2011.co.in/census/state/gujarat.html

Gupta, R. K. (June, 2004). Water governance in Gujarat State, India. *Water Resources Development, 20, 2*, 131–147.

Gupta, R. K. (2011). The role of water technology in development: A case study of Gujarat State, India. Paper presented at the UN-Water International Conference, Zaragaza, Spain.

Hashmi, S. (May 5, 2014). Can India's economy model itself on Gujarat? BBC News. Online at: http://www.bbc.com/news/business-27257790

Hirway, I. (2005). Ensuring drinking water to all: A study in Gujarat. Paper presented at the 4th IWMI-TATA Annual Parterns Research Meet, February 24–26, New Delhi.

IANS. (August 4, 2014). India should follow Gujarat model in farming: Economist. *The Economic Times*.

Iyer, P. (2013). *Gandhi's Gujarat: Vibrant diversity*. Monterey, CA: Center for Conflict Studies.

Kaushik, A. (June 3, 2010). Mahatma Gandhi and enviroment protection. Online at: http://www.mkgandhi.org/articles/environment1.htm

Kishore, A. (June 10, 2014). When wells fail: Farmers' response to groundwater depletion in India. Agriculture and Ecosystems Blog, December 2014. Online at: https://wle.cgiar.org/thrive/2014/06/10/when-wells-fail-farmers'-response-groundwater-depletion-india

Kulkarni, M. (December 3, 2013). Drainage stystem of Harappan Civilization/Indus Valley. Retrieved from http://www.importantindia.com/8962/drainage-system-of-harappan-civilization/

Langa, M. & Gadhavi, D. (April 30, 2014). In pictures: Ahmedabad's segregated Muslims. *Aljazeera*.

Mathur, N. (December 1, 2012). On the Sabarmati riverfront: Urban planning as totalitarian governance in Ahmedabad. *Economic and Political Weekly, 47* (47–48), 01 December 2012.

Mehta, L. (1995–1996). Water, difference, and power: Kutch and the Sardar Sarovar (Narmada) Project. Online at http://www.ids.ac.uk/publication/water-difference-and-power-kutch-and-the-sardar-sarovar-narmada-project

Mehta, L. (2001). The manufature of popular perception of scarcity: Dams and water-related narratives in Gujarat, India. *World Development, 29, 12*, 2025–2041.

Nair, A. (June 19, 2013). Number of farmers in Gujarat dips by 3.5 lakh in 10 years. Retrieved from http://archive.indianexpress.com/news/number-of-farmers-in-gujarat-dips-by-3.5-lakh-in-10-years/1130903/

Narula, K., Fishman, R., Modi, V., & Polycarpou, L. (2011). Addressing the water crisis in Gujarat, India. Online at http://water.columbia.edu/files/2011/11/Gujarat-WP.pdf

Peer, B. (April 18, 2014). Being Muslim under Narendra Modi. *The New York Times*.

Pinto, V. S. (November 12, 2014). Coke's Gujarat plant in troubled waters. *Business Standard*.

Prakash, A. & Sama, R. K. (February 18, 2006). Social undercurrents in a water-scarce village. *Economic and Political Weekly*.

Sachitanand, R. (August 1, 2013). The two faces of Narendra Modi's development model. *The Economic Times*.

Schulte, P. (February 4, 2014). Defining water scarcity, water stress, and water risk: It's not just semantics. Online at: http://pacinst.org/water-definitions/

Soni, J. (2006). Water accessibility and marginalisation of Dalits: Some observation of rural Gujarat. Paper presented at the Water, Law, and the Commons, Delhi.

Subramanian, A. (January 23, 2013). Arvind Subramanian: Is the Modi miracle overrated? *Business Standard*.

Subramanian, T. (2010) "The Rise and Fall of Harappan City"The Archaeology News Network. https://archaeologynewsnetwork.blogspot.in/2010/06/rise-and-fall-of-harappan-city.html

United Nations. (2010). The human right to water and sanitation. Retrieved from http://www.un.org/waterforlifedecade/human_right_to_water.shtml/

Upadhyay, B. (2004). Gender roles and multiple uses of water in North Gujarat. Online at: http://www.iwmi.cgiar.org/Publications/Working_Papers/working/WOR70.pdf

WWAP (United Nations World Water Assessment Programme). (2014). The United Nations World Water Development Report 2014: Water and Energy. Paris, UNESCO. Available online: http://unesdoc.unesco.org/images/0022/002257/225741E.pdf

13 In hidden view

How water became a catalyst for indigenous farmworker resistance in Baja California, Mexico

Marcos López

Introduction

In 1891, the Mexican state chartered a British company to build the trans-peninsular railroad, connecting Baja California's San Quintín Valley with the southwestern United States. The Mexican state's dream was to open this land up for cultivation and to attract investment and settlers (Heath, 2011). The project only produced a 27-kilometer stretch of rail that went nowhere. Due to its proximity to the United States, the Mexican state has long dreamed of rolling out its vision of agrarian modernism, a vision of an agriculture resembling that of its neighbor to the Global North, here in Baja California. Time after time, Mexico has attempted to make this region an extension of the Californian landscape, which, thanks to its agriculture and shipping infrastructure, has become the breadbasket of the United States. However, lack of sufficient water and inhabitants has destabilized this aspiration. The dream of a modernized agriculture in the Mexican state of Baja California, like the dream of the trans-peninsular railroad, has struggled to materialize.

While agrarian modernism has largely failed in Baja California, it persists in the San Quintín Valley, albeit in changing form. Today, U.S. firms in the valley attempt to produce fresh strawberries with the same qualities as those they cultivate elsewhere. Production in Baja California enables American firms to sell strawberries when they are not in season in the United States, but to do so, these firms must simplify the region's ecologies by rationalizing the production process. For instance, firms irrigate by pumping scarce water from coastal water wells, leaving sparse amounts for non-humans and humans in the area. Most recently, these firms have begun to use private reverse osmosis systems to remove salt from ocean water and groundwater, but the dumping of excess salt back into local waters destabilizes the local ecology. This essay examines how the simplification of the regional landscape through export-oriented strawberry production led to the emergence of a new indigenous politics that simultaneously contests unequal access to water and abysmal labor conditions.

This chapter is divided into three parts: the first part introduces the reader to the region's ecological characteristics and builds on what Anna Tsing calls "machines of replication," such as plantations, applying these ideas to my analysis of production and water politics in the San Quintín Valley. The second part explores the way in which capitalist practices have remade the valley into a post-apocalyptic landscape where water is highly politicized. And finally, the third part describes indigenous workers' efforts to remake this landscape by engaging traditional practices rooted in their diaspora. Through this engagement, indigenous workers are building a new kind of solidarity. This essay contributes to discussions on freshwater politics by exploring a newly emerging form of politics sparked by indigenous workers' cultural practices in a region where strawberries are thirsty and water is scarce.

Plantation, replication, and water politics

Plantations are simplified ecological landscapes with the sole purpose of monocultivation. They are also what environmental anthropologist Anna Tsing calls replication machines (n.d.), landscapes devoted to the production of sameness. These landscapes are characterized by the removal of materials and organisms from their life worlds in order to make way for what we have come to call progress. But progress comes at a great price. The concept of machines of replication resonates with the ideas of other scholars who emphasize the material uniqueness of locality within agriculture (Bowen & De Master, 2014; Walker, 2004). A replication machine operates within a historical context that ties the particularities of place to other global connections found within supply chain capitalism (Tsing, 2009). Yet, the replication machine provides insights into local human and non-human entanglements implicated through these connections. Strawberry production and the desertification of the San Quintín Valley, for instance, takes place against the backdrop of neoliberal deregulation and runaway capitalism, where a reversal of agrarian reform, privatization of water, incentives for building infrastructure, foreign investment, new technologies, and indigenous diasporas give rise to an agrarian modernist state that seems to work but is all the while failing.

In the San Quintín Valley, plantation production attempts to replicate strawberries as a universal commodity, while destroying the landscape where the fruit is grown. This replication machine simplifies water and labor as inputs, and thus reflects a way of thinking in which ecologies are separate from humans. Following global trends of privatization (Swyngedouw, 2005), water is simplified as a resource necessary for cultivating fresh strawberries. Once pumped from the ground and treated for irrigation, water is reduced to a utilitarian resource that holds no other meaning or purpose. Likewise, in this system, labor functions as an input, the cost of which lowers south of the United States–Mexico border. In order to make strawberry production more efficient, requiring the least possible amount of time and the fewest possible number of hands, farmworkers, like in other locations (Holmes, 2013), are told how to contort their bodies across the fields, what kind of fruit to pick, and how fast to move. Sometimes this takes place under the pretense of best practices, and sometimes it simply looks like growers blatantly hurrying workers along with a stick or yelling insults.

Though alienated labor practices attempt to reduce workers to inputs, they, like other organisms in the landscape, create obstacles for the replication machine. Through their labor workers create the landscape, as the web of complexity that characterizes this simplified ecology creates a new sort of politics (Guthman, 2011). Unlike common Marxist assumptions that link class consciousness to labor protest, I argue that indigenous workers bring different practices and epistemologies through their migration, often using them to make sense of and transform the landscape in their new surroundings. By doing so, workers cultivate a different kind of politics, one that sometimes combines strange bedfellows, such as contesting inequitable access to water and decrying poor labor conditions.

I first visited the San Quintín Valley in Baja California, Mexico, in 2002. My informants in California's Pajaro Valley, where I began my research on strawberry production, had indicated that firms were beginning to expand production to Baja California. They were looking to ameliorate their economic losses caused by the imminent phaseout of methyl bromide in the United States. Methyl bromide is a pre-plant fumigant used to eliminate soil-borne microbes such as *Verticillium dahliae*, a fungus that causes strawberry plants to wilt. In plantation production such as developed in the Pajaro Valley, this fungus can destroy an entire strawberry crop. Use of methyl bromide, when combined with chloropicrin, enabled California strawberry growers to achieve miraculous annual yields of up to thirty tons per acre. With its mild summer temperatures and proximity to national distribution chains, the Pajaro Valley grows forty percent of all strawberries produced in the United States.

However, in the 1990s, the parties to the Montreal Protocol, which aimed to eliminate ozone-depleting substances, found that methyl bromide adversely affects the global environment, and, in an effort to mitigate climate change, they made plans to phase it out (Mayfield & Norman, 2012).[1] Pesticide firms and the U.S. strawberry industry lobbied to extend the deadline past 2005, while the Protocol gave Mexico until 2015 to phaseout the fumigant because it recognized the country as a developing nation (Carpenter, Gianessi, & Lynch, 2000). With no viable alternatives to methyl bromide, the industry began to expand operations south of the border to the San Quintín Valley, where it found the region's climate amenable to cultivating strawberries.

My preliminary research, however, showed that ecologists, economists, and anthropologists studying the San Quintín Valley predicted that the region's agricultural production was doomed due to decades of commercial agriculture and state dreams of agrarian development that had devastated the local watershed. I wondered how an industry so dependent on water would succeed in this place. In 2003, I drove my 1988 Toyota truck down from Santa Cruz, California, through the Pajaro Valley. I crossed from the lush, manicured cityscapes of Los Angeles and San Diego into an economically and ecologically devastated Tijuana, where my mother grew up before migrating to the United States in the 1960s. Beyond the Bahia de Todos Santos in Ensenada, just an hour's drive south from Tijuana, I discovered a different Baja California. The curved edges of the Sierra de San Pedro Mártir range opened up to the San Quintín Valley, a coastal desert valley that, at first glance, seemed desolate and bare (see Figure 13.1).

Here, I began to conduct ethnographic research on strawberry production in the valley, looking at how migrant farmworkers dependent on seasonal wage work organized amidst their poverty. Over a seven-year period, I interviewed more than fifty informants, from plant scientists to firm managers, farmworkers to government officials. Additionally, I carried out ethnographic research in three indigenous farmworker squatter settlements, called *colonias*, in 2009. Here in these *colonias*, I learned how indigenous workers interpret their labor conditions, how they relate to water, and how they use cultural practices to organize against the state.

The desert landscape

A series of streams originating in the Sierra de San Pedro Mártir range to the east and running down the mountains' western slope intersects on the San Quintín Valley floor. After intersecting, they flow through the valley and out into the Pacific Ocean. While most of these creeks run dry today for the majority of the year, this was not always so. Older residents shared stories with me of bathing and frolicking in these waters each spring, and they spoke of cattle ranchers hydrating their herds at the valley headwaters. The valley's floor once boasted a wide range of wild grasses that browned as the summer progressed.

Settlement has always been challenging in the San Quintín Valley. Although little has been written about it, evidence points to indigenous settlements in this area by the Kiliwa people. Diseases unknown to these peoples and brought by Dominican missionaries, who established a mission at the base of the mountains in 1775, wiped out local populations. Twenty years after establishing the mission, water became too scarce to support human life, and the Dominicans left abruptly (Vanderplank, 2011, p. 65). In 1883, Mexico's liberal government granted a charter to English and American colonists, hoping they would populate the region and bring agrarian modernism to a portion of the nation known to attract American filibustering (Velasco, Zlolniski, & Coubes, 2014, p. 66). Colonists set up a wheat plantation and constructed a gristmill, but they abandoned their project when drought made production and their livelihoods untenable.

During my interviews, older residents also recalled hearing stories of once abundant sea otters preying on shellfish in the shoreline's bays and estuaries. Overhunted, sea otters went extinct in the area during the fur trade, which lasted through the 1850s. Today, none of these playful creatures remains, though residents nostalgically imagine the day when sea otters will return to the valley. Since many of the area's wetlands have now dried out and are unable to support such life, sea otters will never actually return, despite residents' hopeful sentiments.

In the early 1980s, export-oriented agricultural firms moved into the San Quintín Valley and triggered a process of desertification (Schwartz & Notini, 1994, p. 8). Growing tomatoes for U.S. consumption, American and Mexican

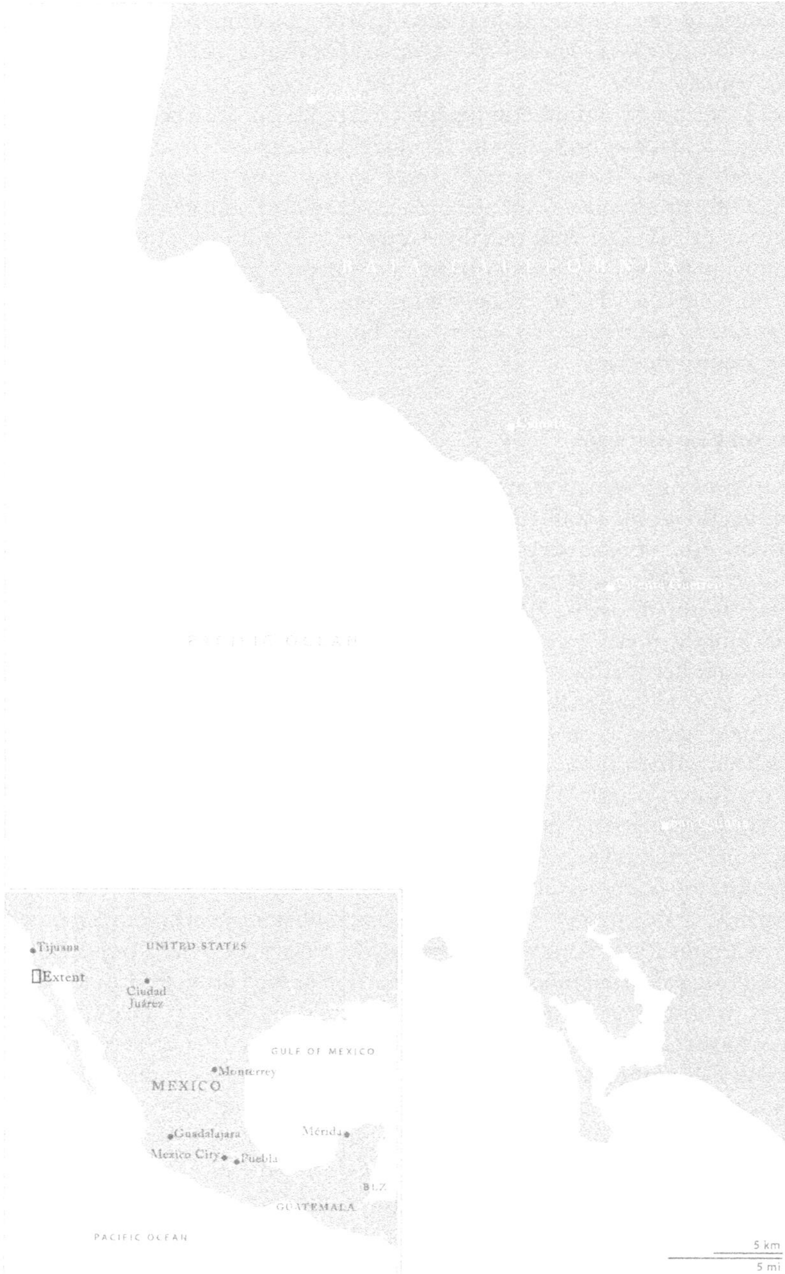

Figure 13.1 San Quintín Valley, Mexico

Source: Cartography by Gregory Woolston and Bill de la Rosa using data from Natural Earth and © OpenStreetMap Contributors.

firms mined the valley for fresh water by digging new wells, wreaking havoc on the local water supply. Extracting water from low-lying coastal aquifers is problematic because once the wells are overdrafted, dense seawater seeps through the coastline's sandy loams and contaminates the aquifer, ruining the local supply of fresh water. This irreversible process has transformed the San Quintín Valley watershed into a pool of brackish water. In most of the valley, salinity levels reach between 3000 and 9000 parts per million (ppm) (Aguirre-Muñoz et al., 2001; Zlolniski, 2011), making the water useless for sustaining human life. In 2002, prominent agricultural economist Roberta Cook stated that, due to ruination of the watershed through saltwater intrusion, commercial agriculture in the San Quintín Valley had no future. The irony here is that where the Mexican state sought to create modern agrarian development, it instead destabilized ecologies to the point of collapse (Vanderplank, 2011).

The "art" of making water

When, prompted by the methyl bromide phaseout, I first arrived in the San Quintín Valley in 2002 to study strawberry production expansion, I witnessed an alternative future unfolding. Strawberries—a water-intensive crop—were appearing throughout the valley, followed by raspberries, blueberries, and the region's already existing abundance of tomatoes. I wondered: Why were agricultural firms in the valley radically expanding production when experts claimed water contamination would derail the industry? Growers whom I interviewed offered a quick answer: Desalination plants were creating new possibilities for a place previously thought to have a bleak future. As one grower boasted, "It costs us about one [U.S.] dollar to produce one cubic foot of water that we can use for irrigation." Today, agricultural firms produce water with twenty-five privately-owned desalination plants that remove salt from the brackish aquifer reserves remaining in the valley (Zlolniski, 2011, p. 572). SAGARPA, Mexico's Department of Agriculture, heavily subsidizes the building and operating of desalination plants and well-water pumps. SAGARPA subsidizes up to fifty percent, or 750,000 pesos, of the cost to build a desalination plant for commercial agriculture, while they give firms an additional 0.33 pesos per kilowatt to reduce the electrical costs of operating their desalination plants and well-water pumps (Facchini, 2009, p. 24).

Water desalination is a complex and cost-intensive venture. Agricultural firms compete to locate water deposits across the San Quintín Valley. Though Mexico's National Water Commission (CNA), which regulates the use of water for commercial agriculture in Baja California, no longer issues licenses for new water wells because of the valley's chronic overdraft problem, the region's largest and most powerful agricultural firms nonetheless mine new water wells. Mexico's decentralized water policy requires that stakeholders manage local watersheds (Wilder & Lankao, 2006, p. 1978). The San Quintín Valley contains six aquifers, each governed by its own committee, but these committees are composed of representatives from the valley's largest agricultural firms

(Zlolniski, 2011, p. 574). As a result, self-regulated agricultural firms drill illicit wells on their property, while others divert water at the headwaters of the water basin, some 23 to 43 kilometers from the valley floor (Zlolniski, 2011, p. 572).

The limits of turning brackish water into irrigation water

In 2009, I began to hear predictions about the future of commercial agriculture in the San Quintín Valley. Surprisingly, the complaints came from growers themselves. During an interview at a farm growing strawberries for Andrew and Williamson, an American firm selling in the United States under the label Limited Edition, Miguel, the mestizo manager hired by the Mexican grower to oversee production, described instabilities produced by water scarcity.[2] In 2008, the grower had planted a 101-acre field of strawberries to be harvested from mid-January through June. However, Miguel explained that by early May, the well supplying water to this field had run dry, causing the entire crop to die before the harvest season ended. The following year, the grower located a new well across the highway from his old field. Engineers developed a reverse osmosis system with the goal to cultivate 124 new acres of strawberries. Andrew and Williamson brokered a deal for the grower to receive the credit necessary to build the new water infrastructure he needed. Though expanding production by adding new fields and constructing more desalination plants compensates for the instability created by overdrafting the local aquifer, the technological fix never completely achieves its goal, as nature continues to create obstacles for capitalist expansion.

In 2012, agricultural firms extracted 47 million cubic meters of groundwater, while the annual recharge is about 35 million cubic meters. In a region that receives 15 centimeters of annual rainfall, approximate annual water loss ranges from 10 to 12 million cubic meters (Martinez Veloz, 2012). Some of the annual recharge does not come from rainfall, but from saltwater intrusion. One berry grower told me that the groundwater supply is quickly vanishing and that growers are scrambling to find other places where they can extract water. He said, "One grower already pumps water directly from the ocean. It's only a matter of time before we all need to do the same." While walking on the beach one day, I came across the intake this particular grower uses to gather water from the ocean: a long flexible pipe, perhaps two meters in diameter, which runs across the coastline and into the depths of the ocean. I can only imagine what a shoreline crowded with intake pipes would look like in the near future.

Since strawberry plants are inherently thirsty, requiring twenty gallons of water to produce one pound of fruit, their production in the San Quintín Valley is looking bleak. While certain plants such as mangroves can thrive in brackish water, strawberry plants lack the ability to process excess sodium. Accumulation of sodium stresses the plant by limiting photosynthetic efficiency and root development (Kaya, Ak, & Higgs, 2003), thus ruining the harvest. The plant's shallow roots require constant watering in order to keep the plant alive and make it productive.

In order to use brackish water for large-scale production, valley growers must first lower salinity levels to 550 ppm. Firms cluster a series of registered and unregistered wells, normally three or four, where they pump groundwater and pipe it into a company-owned reverse osmosis system. There, they blast water through a series of cylindrical vessels filled with synthetic membranes that filter out salt (Wilf, 2014, p. 157). Higher salinity counts require two gallons of seawater to produce one gallon of irrigation water. While this process does not make water suitable for human consumption (50 to 300 ppm), it does sufficiently lower salinity levels for the cultivation of strawberries, tomatoes, raspberries and blueberries.

Humans competing with plants for water

One of the San Quintín Valley's greatest ironies is that the very people who pick strawberries for the firms must also compete with them for water. The Mexican government has made the valley's commercial agriculture economically viable by subsidizing the cost for producing water and turning a blind eye to illegal water mining. Yet the same water merely trickles into the settlements that house the farmworkers who make the harvest possible. The majority of the valley's *colonias* receive weekly water deliveries from CESPE, the municipal water agency. Each week, CESPE's water trucks circulate through farmworker settlements and, for 15 pesos, fill one 200-liter blue plastic barrel for each household in the *colonia*. Two hundred liters equate roughly to 60 gallons of water, the amount required to grow three pounds of strawberries. However, this amount never actually meets household needs, as there are often multiple families living in a single household. In Colonia Morelos, Leticia, a Mixtec woman, told me that her family consumes an additional 400 liters of water per week. She must be vigilant to flag down private vendors to fill the additional two barrels she uses to get through the week. The water she buys is still too salinated for human consumption, and Leticia's family uses it only to wash clothes, water plants, and flush toilets. She must also purchase drinking water from private venders at 12 pesos for a five-gallon bottle.

Other farmworker settlements pipe in water. Typically, whether through protest or patronage, these *colonias* have forced the hand of the state to install modern water infrastructure. Yet farmworker households assume responsibility for transporting water from the main water line to their homes, paying the meter that monitors their consumption. Rather than running water throughout the home, most households rely on a spigot at the edge of their property, which can be turned on and off as needed. Adding to the precarious nature of these services is the fact that the state strictly rations all water distributed to *colonias*. According to farmworkers living in these settlements, the rationing method is far from equitable. Sometimes, without notice, the state turns off the valve for days at a time, leaving communities without water.

Informants I interviewed pointed out that the line between legal and illegal land ownership in the San Quintín Valley is blurred, which, they claim, state agencies use to deny indigenous farmworkers access to water. Changes to

Article 27 of the Mexican Constitution, which regulated land ownership at the federal level, gave valley agricultural firms the opportunity to expand operations and shutter labor camps, while other landowners urbanized their farmland. Displaced indigenous farmworkers sometimes came together to collectively repurpose abandoned farmland as squatter settlements, while others became victims of land fraud. Landowners exploited an absence of state oversight to sell the same plots several times over. Today, lack of proper land titles and knowledge of the bureaucratic process renders indigenous people illegal and invisible in the eyes of the state. In an interview, a lawyer hired by the state to defend the rights of indigenous people spoke in support of this logic: "Why should the state put a water line in a *colonia* that should not have existed in the first place?"

In the San Quintín Valley, the machine of replication works through the tools of modernity whereby the state mobilizes nature, people, and capital, while at the same time leaving indigenous workers out of what has been called progress. State documents are one mechanism by which farmworkers are rendered invisible. Birth certificates, land titles, social security cards, and other documents make citizens legible to the state, threading them into a discourse of legality and illegality. Because indigenous people are frequently born outside of state institutions, they often lack the paperwork needed to make them visible even in their own nation. Without birth certificates, many indigenous people are unqualified to vote, receive medical services and state benefits, cash paychecks, and legally own land. On these grounds, agricultural firms as well as the state deny indigenous farmworkers access to basic services such as water, state benefits, and protection from abuses, including unsafe working environments. For instance, many growers I interviewed claimed they were unable to pay into mandatory social security and health insurance benefits or issue paychecks. Therefore, paying farmworkers in cash is common practice, making them vulnerable to abuse.

In the absence of official paperwork, firms are at liberty to hire indigenous people as temporary workers for a single day—or even simply for part of it—and fire them as needed. Farmworkers are often required to work seven days per week, and since workers cannot prove their seniority or employment status, nor complain to the labor secretariat's office, more insidious abuses ensue. Workers I interviewed regularly complained of pesticide exposure and mistreatment by supervisors, including sexual harassment and rape. Some reported that growers regularly hire off-duty police officers or armed security to patrol the fields and remove unruly workers.

Diaspora and resettlement: landscape, practices, and protest

While water is one of the most costly and critical resources needed to keep thirsty strawberry plants alive, the capitalist project of agrarian modernism in the San Quintín Valley cannot survive without workers to pick the fruit. Indigenous groups from the Mexican state of Oaxaca have become the preferred workers for the firms operating in this coastal region. But why indigenous workers, and how have these workers reached the valley?

The San Quintín Valley was an isolated region until the 1970s when, with the completion of the trans-peninsular highway connecting Baja California with markets in the United States, firms began seeing possibilities for developing export agriculture. In order to find farmworkers, U.S. and Mexican firms sent labor contractors to rural Oaxaca, who told indigenous people that well-paying jobs and free housing awaited them in the valley. Instead, however, new workers unknowingly arrived to jobs paying poverty wages and decrepit housing in firm-operated labor camps (Garduño, Garcia, & Moran, 1989; Nagengast & Kearney, 1990). The cycle of poverty and debt kept them from returning home. The passage of the North American Free Trade Agreement in 1994 exacerbated the exodus of indigenous migrants. When they could no longer compete in the market to sell the crops they grew in Oaxaca, indigenous people migrated north. At the same time, valley firms saw a production boom when the introduction of strawberries made agriculture a year-round venture requiring even more workers.

Because of labor migration, the region's population rose from 38,151 to 92,177 between 1990 and 2010 (Coubes, Velasco, & Zlolniski, 2014, p. 20). Migrant settlement, even under precarious circumstances, has become common. Depending on the history of each *colonia*, Triqui and Mixtec or even Zapotec and Nahuatl people—groups hailing from different territories, often with a history of friction in their homelands—commonly live in the same settlement.

At his home in one of the valley's many *colonias*, I met Lucio—one of my first informants and one of the most well-known indigenous leaders still alive. Dark, stocky, and self-identified as Nahuatl but with strong Triqui affiliations, Lucio had been organizing indigenous people to fight labor abuses in the valley since the early 1980s. He explained to me that after various attempts on his life and the assassination of his brother, also an organizer, Lucio was no longer able to find work in the fields: all the growers had blacklisted him. Committed to staying in this community, Lucio became a self-employed fisherman gathering large Pismo clams by night and a leader by day.

In a repressive environment where leaders are often driven out of the community or mysteriously killed, Lucio had grown skeptical of researchers, union organizers, firm representatives, government officials, and other outsiders like myself. In fact, community members first thought I might be an informant hired by the firms to infiltrate. It took many days and hours standing at a dusty corner near Lucio's home asking unwilling farmworkers simple but important questions about pesticide exposure and labor conditions to convince him that my partner and I were benign enough to trust. Lucio invited us to learn how different indigenous groups make new lives for themselves in these *colonias*. He encouraged us to attend community activities and see how people live.

One of the most important practices bringing these different indigenous groups together is a public gathering known as the *asamblea.* On Sundays, often the only day off for valley farmworkers, a group gathers in the morning at a dead-end street in Colonia Las Lomas de San Ramon. At most gatherings, Lucio opens a dialogue where residents typically ask him pressing questions such as: When will the state turn the water back on? Why has a neighboring *colonia* received an additional water line while we wait in limbo? Lucio delivers to residents the responses he has received from state officials, and the group discusses which actions to take. After one particular *asamblea* I attended, Lucio told me that if I truly wanted to learn about the *colonia*'s struggles, I would not only need to observe *asambleas*, but participate in the community's activities.

That Sunday after the *asamblea*, we gathered to fulfill a *tequio*, a sort of collective work that involves the entire *colonia*. The goal was to ready the terrain for a new road, with the expectation that the state would pay with their own *colaboración* (collaboration) by laying down the cement. *Colaboración* is a common practice among groups from Oaxaca and Guerrero: indigenous people raise money or collectively offer their labor in order to see a community project come to fruition. The group saw this road preparation as a *colaboración*, where the state too would do its part. With spades and shovels we removed weeds and roots from the soil. I watched children, mothers, elders, men, and women all contribute to the work. The mood was playful, and people sang and teased each other, yet even the little ones worked earnestly to clear the way for the road. Although they have not always met with success, leaders have learned to use *tequios* and *colaboraciónes* to compel the state to cooperate in community projects.

Something else was taking place on this Sunday morning as well. Leaders were following up on the progress of a previous *tequio*, where residents had come together to green up the street by planting palms and trees. Calling out the names of particular residents followed by the plants for which they were responsible, the voice on the megaphone summoned the attention of those whose plants had died due to lack of water or care. To fail the plant was to fail the community's effort to remake this landscape into something remotely resembling a nature these people once knew in Oaxaca or Guerrero. Though some participants complained that they were unable to keep their plants alive due to salinated water, they had committed to making them thrive on these arid and dusty streets and willingly took on the responsibility of new ones in another attempt to enliven this place. Although I saw many anemic-looking palms, they were tenderly protected from gusty winds with plastic and frequently watered with the valuable little water at the community's disposal. Witnessing this insistence on transforming their post-apocalyptic landscape into something resembling their idea of nature in southern Mexico, despite constant failure due to the lack of fresh water, helped me to understand, indeed experience, the emergence of anger and sadness in these communities.

The landscape in Oaxaca produces a particular affect for indigenous migrants. Andres, a Triqui farmworker, sat back, looked down at his hands, and, with a frustrated voice, told me, "You see, in Oaxaca we did not need water. Water was always around us. We could gather it from the streams. It was clean and tasted fresh." Water in the Mixteca Alta, from where the majority of the valley's indigenous farmworkers migrate, is not easy to forget, as its abundance produces much life. Not only does water in this region give energy to human life, but also to *milpa,* pine forests, and livestock—life that struggles to exist in the San Quintín Valley.[3] In Oaxaca, where water is not commodified, human relationships with nature produce different ways of understanding water.

In Oaxaca and Guerrero, water and other elements of nature are not dominated, but understood, revered, and longed for when no longer present. Maria, a Mixtec farmworker in the valley, echoed this sentiment: "We cannot harm what gives us life, we must show it respect." Cut off from most of southern Mexico, the Mixteca Alta is a mountainous landscape, where, for centuries, people foraged and practiced subsistence agriculture. Its steep slopes and deep valleys provide a combination of volcanic and alluvial soils that generate livelihoods connected to the particular ecologies. Moreover, people use local knowledge to cultivate subsistence crops, such as beans, maize, and squash, grown together and known as *milpa*. In the Mixteca Alta, wind, maize, earth, and water form a complex web of life, a materiality that Mixtecs and Triquis view as inseparable (Joyce & Gorman, 2012, p. 3). In this region, maize is nothing without the energy provided by the sun, while rainfall gives life to streams and the harvest.

Though the Mixteca Alta is located far from the San Quintín Valley, indigenous people continue to remember the life they left behind. Remembering produces both sadness and anger. Estella, a Mixtec farmworker who arrived to the San Quintín Valley in the early 2000s, migrated from Oaxaca convinced that a job in the fields would improve her life. She travelled by bus with her parents for three consecutive days, arriving at her new home in the middle of the night. As morning broke, she hurried outdoors to see the landscape that awaited her:

> We arrived at night to a *cuarteria* [farmworker cinder block housing]. In the morning I was shocked by what I saw [a barren desert], and I cried when I saw my new living conditions. I could not believe that I was now living in the desert with nothing.

What Estella encountered in the valley was not natural to her. She saw an unwelcoming and empty landscape.

Because water is scarce in the San Quintín Valley, the relationship with nature that Estella and other indigenous people enjoyed in the Mixteca Alta cannot be reproduced. The desertified landscape breeds different kinds of relations. However, indigenous people have refused to accept the conditions forced on them in this new environment. Telma, an older Triqui farmworker, used graywater collected from washing clothes to water her herb garden. She tested varieties of seeds she brought with her from the Mixteca Alta to see which could withstand their new environment. She sang praises to the herbs that flourished. Adolfo, another Triqui farmworker, explained that successfully growing subsistence crops in his *colonia* is difficult:

> Many people try to grow *milpa* in the *colonia*. It struggles to grow, and most of the time it dies because the water we get [at home from CESPE] all of the sudden will get saltier. I suspect that they [CESPE] are giving us the saltier water and letting the firms have the better water.

Indigenous farmworkers' relationship with nature in the Mixteca Alta engenders a different kind of solidarity in the San Quintín Valley, one that engages relationships they left behind with the materialities and practices they encounter in this broken landscape. Sadness and anger bring people together through the indigenous practices of *asambleas*, *tequio*, and *colaboración* in an effort to enliven the place. From these practices, new solidarities, new collectivities, and a different kind of indigenous politics have emerged, presenting serious challenges to the project of agrarian modernism in the San Quintín Valley.

In 2009, indigenous solidarities materialized when valley communities began organizing around increased water rates. When the state increased water rates by 4 pesos, community members argued that the state was not honoring their *colaboración* of 52.50 pesos. Indigenous groups had, in their view, supported the work of state officials with their *colaboraciónes* by paying taxes, but the state had not done its part in consulting indigenous settlements about charging more for water. In their eyes, state officials were failing them as leaders. Indigenous groups demanded that their water rates be kept once and for all at the previously agreed amount of 52.50 pesos—a symbolic amount, as this is a farmworker's minimum daily wage (approximately US$3.50). At first, state officials ignored these demands, claiming that inflation rates had increased the cost of water services, but indigenous people were insistent. They demanded to meet with state officials, and when their demands went unmet, indigenous groups refused to pay their water bills, staging a massive sit-in in front of CESPE's regional office. With funds collected among community members, indigenous leaders traveled three hours to Ensenada to speak with CESPE's municipal director, where they explained that the state should not institute such increases without first consulting the community.

Eventually, CESPE's officials entered the *colonias*, where, on a blustery spring evening they met with indigenous people from a handful of farmworker settlements. Triqui women wore their *huipiles* as they walked in groups across the *colonia*, painting the landscape with streaks of red, pink, green, and orange. Women, children, and men of all ages attended this large *asamblea*. Standing at the center of the *colonia*'s plaza, state officials apologized for not consulting the community and promised to keep water at the previously agreed upon rate. The state kept their word and promised to renegotiate the following year.

This new solidarity gave birth to the *Comité de Colonias* (the Committee of Colonias). Lucio spearheaded the initiative by working collectively with other indigenous leaders when there were decisions to make. This collectivity enabled *colonias* to act in solidarity when negotiating with the state. In March 2015, the *Comité de Colonias* organized the valley's first successful general strike, which shut down agriculture in the region for one month. The strike ended when the state, not the firms, agreed to subsidize wages by an increase of 200 percent. For indigenous leaders, this accomplishment was the fulfillment of a dream to organize their communities in contestation of labor violations in the fields, including wage theft, sexual violence, pesticide exposure, humiliations, and unfair firing practices—and to make the *colonias* a place resembling the life they had left behind in Oaxaca and Guerrero.

Conclusion

Machines of replication are capitalist projects that make nonsensical decisions about ecologies, ecologies that include both humans and non-humans. The destructive growing of plantation strawberries in the San Quintín Valley, which simplifies the landscape by extracting salt from water and reducing workers to rationalized inputs, is an instructive example of a machine of replication. In this case, the Mexican state orchestrates development using modern tools that engender coercion, violence, and the erasure of landscapes. However, simplified ecologies have their limits. When will the valley's overextracted aquifer finally run dry? And when it does, what will it look like to pump massive amounts of seawater from the ocean? What happens to the excesses that emerge, such as salt and aquatic life? And, with regard to farmworkers, how far will the state go to permit labor abuses in light of its attempts to attract capitalists that can carry out its dream of agrarian modernism?

Indigenous people understand water differently than assumed in western logics. Rather than seeing it as a resource that needs simplification and disentangling, they revere water because of its complexity and its entanglements with the landscape. Water, after all, is what gives life. However, in the San Quintín Valley, the water that gives life to strawberries does a poor job of giving life to farmworkers. The simplification of water in the valley presents a challenge to indigenous people. The nature they attempt to cultivate cannot be achieved in this destroyed landscape. From their sadness and anger emerges a new solidarity, and with it, a vision for a new indigenous future that operates outside the logic of the state and its destructive development dreams.

Notes

1 Established in 1987 and put into effect in 1989, the Montreal Protocol warned of a widening hole in the ozone layer. The Protocol set in place a multilateral agreement for phasing out ozone-depleting substances such as methyl bromide.
2 To maintain confidentiality, I have assigned pseudonyms to all interviewees.
3 *Milpa* is a small-scale, integrated crop-management system used extensively throughout Mesoamerica.

Works cited

Aguirre-Muñoz, A., Buddemeier, R., Camacho-Ibar, V., Carriquiry, J., Ibarra-Obando, S., Massey, B. W., Smith, S.V., & Wulff, F. (2001). Sustainability of coastal resource use in San Quintin, Mexico. *Ambio, 30*, 3, 142–149.

Bowen, S., & De Master, K. (2014). Wisconsin's "Happy Cows"? Articulating heritage and territory as new dimensions of locality. *Agriculture and Human Values, 31*, 4, 549–562.

Carpenter, J., Gianessi, L., & Lynch, L. (2000). *The economic impact of the scheduled U.S. phaseout of methyl bromide.* Washington, DC: National Center for Food and Agricultural Policy.

Cook, R. (2002). Strawberry production in the United States, 1990–2000. Retrieved from http://ucanr.edu/datastoreFiles/234-27.pdf.

Coubes, M. L., Velasco, L., & Zlolniski, C. (2014). Residential settlement and mobility in the San Quintin Valley: Methodological reflections on an interdisciplinary study. In L. Rivera-Sánchez & F. Lozano-Ascencio (Eds.), *The practice of research on migration and mobilities* (pp. 19–41). New York: Springer.

Facchini, G. (2009). *Water scarcity in Mexico*. Oslo: University of Oslo.

Garduño, E., Garcia, E., & Moran, P. (1989). *Mixtecos en Baja California: El caso de San Quintin*. Mexicali, Mexico: Universidad Autonoma de Baja California.

Guthman, J. (2011). *Weighing in: Obesity, food justice, and the limits of capitalism*. Berkeley, CA: University of California Press.

Heath, H. (2011). El malogrado proyecto del ferrocarril peninsular del Districto Norte de la Baja Califonria, 1887–1892. *Estudios Fronterizos, 12*, 2, 185–221.

Holmes, S. (2013). *Fresh fruit, broken bodies: Migrant farmworkers in the United States*. Berkeley, CA: University of California Press.

Joyce, A., & Gorman, M. (2012). Bridging the theoretical divide in halocene landscape studies: Social and ecological approaches to ancient Oaxacan landscapes. *Quaternary Science Reviews, 55*, 1, 1–22.

Kaya, C., Ak, B. E., & Higgs, D. (2003). Response to salt-stressed strawberry plants to supplementary calcium nitrate and/or potassium nitrate. *Journal of Plant Nutrition, 26*, 3, 543–560.

Martinez Veloz, J. (2012). La injusta distribución del agua en San Quintín. *La Jornada*. Mexico City. Retrieved from: http://www.jornada.unam.mx/2012/01/27/opinion/023a2pol

Mayfield, E., & Norman, C. (2012). Moving away from methyl bromide: Political economy of pesticide use in strawberries since 2004. *Journal of Environmental Management, 106*, 93–101.

Nagengast, C., & Kearney, M. (1990). Mixtec ethnicity: Social identity, political consciousness, and political activism. *Latin American Research Review, 25*, 2, 61–91.

Schwartz, M. L., & Notini, J. (1994). *Desertification and migration: Mexico and the United States*. Washington, DC: US Commission on Immigration Reform.

Swyngedouw, E. (2005). Dispossessing H2O: The contested terrain of water privatization. *Capitalism, Nature, Socialism, 16*, 1, 81–98.

Tsing, A. (n.d.). Lecture: Earth stalked by man. University of California, Santa Cruz and Aarhus University.

Tsing, A. (2009). Supply chains and the human condition. *Rethinking Marxism, 21*, 2, 148–176.

Vanderplank, S. (2011). The flora of greater San Quintín, Baja California, Mexico. *Aliso: A Journal of Systemic and Evolutionary Botany, 29*, 2, 65–106.

Velasco, L., Zlolniski, C., & Coubes, M. (2014). *De jornaleros a colonos: Residencia, trabajo, e indentidad en el valle de San Quintín*. Tijuana: COLEF.

Walker, R. (2004). *The conquest of bread: 150 years of agribusiness in California*. New York: The New Press.

Wilder, M., & Lankao, P. R. (2006). Paradoxes of decentralization: Water reform and social implications in Mexico. *World Development, 34*, 11, 1977–1995.

Wilf, M. (2014). The reverse osmosis process. In J. Kucera (Ed.), *Desalination: Water from water* (pp. 157–204). Hoboken, NJ: Wiley.

Zlolniski, C. (2011). Water flowing north of the border: Export agriculture and water politics in a rural comunity in Baja California. *Cultural Anthropology, 26*, 4, 565–588.

14 From management to governance

Rethinking water policy and privatization on Easter Island

Maria Alessandra Woolson

Rapa Nui is one of the most isolated places on earth. Located in the South Pacific roughly 3,700 kilometers from South America's coast and 2,112 kilometers from the nearest populated area, the Pitcairn Islands (Fischer, 2010), this island under the political sovereignty of Chile is known in Continental Chile by its Polynesian name and elsewhere as Easter Island. Rapa Nui's remoteness, history, and culture make it a unique setting in which to observe water in its ecological, institutional, and cultural contexts. And yet, while the island's specific set of circumstances makes it unique, it also offers a lens through which to gain broader insights about water politics in Latin America and the Global South.

Rapa Nui consists of three coalesced volcanoes with highly permeable soils. The only source of fresh water available for consumption is an aquifer that sits, like a cushion, under the entire surface of this porous mound. Fed solely by limited rainfall, the quality of this aquifer faces unprecedented risks due to accelerated environmental degradation driven by increased sources of pollution. Human activity is largely responsible for these ecological changes, resulting primarily from accelerated growth in tourism and a growing transient population related to the tourist sector. This recent growth has become disproportionate to the island's carrying capacity and has been enabled, in part, by a Chilean national policy rooted in free market economics that drives a model of development focused on economic growth at the expense of other values. This model of economic growth has pushed this fragile island's vulnerable ecosystem to the point of risking its collapse (Calderón-Azorar & O'Ryan, 2011; Delaune, 2012; ECOPOLIS, 2010; Figueroa & Rotarou, 2013; Rivera, 2012).

Notwithstanding its geographical isolation and particular cultural and ecological context, the Rapa Nui experience raises questions about the suitability of the Chilean approach to water policy with its emphasis on water markets as the key to addressing water access, quality, and quantity. The Rapa Nui experience also highlights some of the problems of managing water as a universal set of technical issues best addressed by the private sector, regardless of context. In its isolation, the island can be understood as a quasi-closed system, a manageable microcosm ideal for observing tensions between divergent approaches to natural resource use: one approach understands natural resources within their cultural and ecological context, while the other understands these resources within development models assumed to be valid regardless of context. This contrast carries serious implications for general principles of accepted best practices for sustaining water resources. For example, as a Special Territory of Chile, Rapa Nui has been subjected to the Chilean "model" of managing water as an economic good, upheld by powerful international organizations as an example of efficiency. But Rapa Nui also features a living culture rich in aesthetic expression, rituals, and ancestral knowledge having to do with the life of water within the life of the community and vice versa—and its multiple connections of water and community are evidence that water is inherently complex.

In addition to referring to the island, the pan-Polynesian term *Rapanui* also refers to the island's ethnic community. As an original people, they have shown extraordinary resilience and adaptive capacity in the face of pressures and transformation. Yet, historically, their cultural expression and traditional practices have been silenced to the point that even some contemporary intellectual and academic communities refer to them as culturally extinct (Haun, 2008; Hunt & Lipo, 2008; Young, 2011). Rapanui history and tradition first became known in the twentieth century by means of transcribed and translated texts, a process mediated in Spanish, the *lingua franca* of the colonizing state that reinforced the community's boundaries and limitations. Some of the islanders' Polynesian beliefs have been inconsistently represented at best. Colonial structures of exclusion decimated the population over several decades and violated even the most basic rights of the island's inhabitants.

Two effects of the community's colonial history are relevant to the sustainability of the island's resource base and fresh water supply, offering helpful insights. One effect is a historic loss of cultural continuity for Rapanui identity, knowledge, and practices—a problem aggravated today by pressures of the market, state, and broader international community. The second effect is the Rapanui's lack of full citizenship rights, which, as we will see, undermines their ability to exercise their rights to their land, to self-representation, and to active participation in the governance of their vital water resource. Human life in Rapa Nui is ultimately defined by the overall quality of its only aquifer, now controlled by a single private operator.

Within the unique context of Rapa Nui, how does valuing water as an economic good respond to both the complex nature of water and the diverse allocation of water rights, including the human right to water?[1] How do more generalized arguments about water scarcity and management efficiency, such as those advanced by Chilean policy, apply to this particular setting? And how do these questions relate to concepts of citizenship and democracy as dynamic and adaptive aspects of water governance, providing opportunities to rethink management in contextually productive ways? In other words, what can we learn from looking at Rapa Nui as an opportunity to reflect on the place-based nature of water problems and solutions where water governance can be productively contrasted with technical water management?

The response to these questions offered in the following pages is part of a larger study of the island's biocultural diversity and results from fieldwork completed in 2010 and 2012. In addition to photographic documentation and a thorough review of the literature published to date, this study relies on ethnographic fieldwork in the form of semi-structured interviews conducted with various stakeholders, program administrators, park rangers, and officials. The first section of this chapter elucidates the socio-cultural significance of this territory and its water resource in order to illustrate the inherent complexity of water for multiethnic societies with a colonial past. The second section contextualizes the topic of water management within the Chilean institutional setting, modeled on market economics and the broader context of international lending institutions. Finally, an analytical composite of recent data and events on Rapa Nui concludes that water governance is the most adaptive process for addressing water quality, quantity, and access, while facilitating the integration of diverse political and cultural perspectives.

Rupturing the bond between people and place

What can looking closely at Rapa Nui teach us about the place-based nature of water problems and the significance of contrasting water governance with water management? In order to address this question, it is important to explore details of culture and practice to draw a contrast. The first important observation to make is that, in spite of its Chilean geopolitical jurisdiction, this community is the easternmost example of Polynesian culture and that is drastically different from other cultural contexts in continental Chile (Woolson, 2014).

Historically, Rapanui territorial organization was based on kinship, specifically structured around the trinity of Polynesian rank and status: nobles, experts, and commoners (Fischer, 2010). Over time, kinship lines ramified, forming large groups distributed throughout the island. High-ranking descendants inhabited the northern and western coasts, where fishing was abundant, while the rest of the population loosely occupied the east and south, developing intensive agriculture. The island was sparsely populated, and collective ownership, anchored in socio-spiritual power derived from the ancestors—or *Mana*—stemmed from these relations. What remains of this historical distribution today is the meaning the Rapanui assign to the land: here, kinship ascribes space its own distinct memory as a cartography that engages the ancestors in dialogue. With this approach to the land, the Rapanui not only cultivate a strong sense of connection with their land and ancestors but, in doing so, also address their anticipated progeny (Young, 2011).

For Rapanui, cosmology and environment—land, ocean, and geophysical elements, including fresh water—are integrally entwined. Their notion of *place* is defined by their relationship to land and resources. Over generations, fresh water and its surface scarcity have shaped many of the community's habits and traditions, including the population's socio-spatial distribution throughout the island (see Figure 14.1). Life for the Rapanui occurred mostly outdoors, and

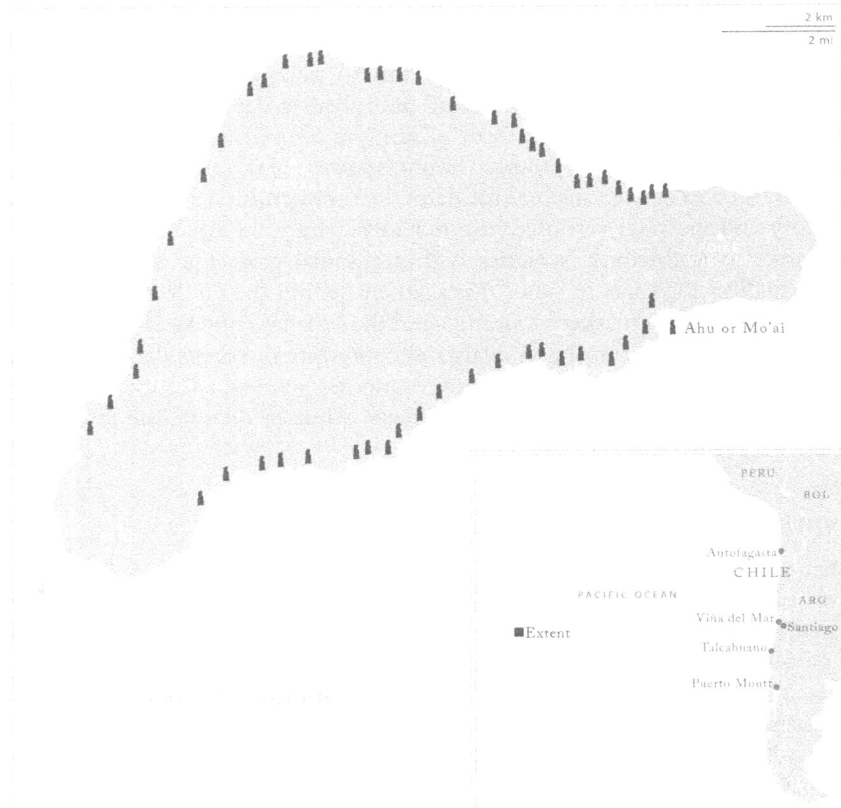

Figure 14.1 Map of Rapa Nui/Easter Island. This map illustrates the location of ahu (ceremonial stone platforms) and mo'ai (monolithic statues), representative of the socio-spatial distribution of the Rapanui population prior to the nineteenth century

Source: Cartography by Gregory Woolston using data from Cooprogetti (2005), Natural Earth, and © OpenStreetMap Contributors.

the house, or the many caves used for occasional shelter, was a place to sleep at night, offering few other comforts.

Water not only shaped habits and traditions but figures prominently in the Rapanui creation story. Like other Polynesians, the Rapanui's relationship with their ancestral deities combined what in Western terms would be the secular and the sacred (Fischer, 2010). Beliefs did not originate in any form of worship. They had a higher socio-spiritual power, or *Mana*, protected by *Tapu*, a ritual elevated to sacred function, which outlined rules governing the use of land, crops, buildings, human behaviors, and beliefs. *Mana* and *Tapu* were interconnected, holding both spiritual and political implications. Within this cosmology, which hinges on the connection between cosmos and material nature, water is life. Nature—humans, the earth, and objects bearing *Mana*—originates with the gods in a merging of earth and sky. Past generations saw water in the form of rain as representing the tears of the deity in the sky who cried, longing for proximity to the Earth God. Fog was also representational: it was the warmth of the Earth God's body longing to embrace its companion. As evidenced by these ancestral beliefs, water has been constitutive of both natural cycles and a spiritual life cycle in Rapanui tradition.

Early Rapanui harvested water from the island's three volcanic craters, fed by seasonal rivers that rushed down the *Avas*, or gullies, during wet winters, as well as from built *Tahetas*, or rock depressions. Many of these depressions bear a leaf-like fertility figure carved on the rock—a powerful socio-spiritual symbol of connectedness between the living and their ancestors. With the historical silencing of such knowledge and practices and with the disruption of these powerful place-based bonds, the Rapanui are no longer able to exercise many of their remaining spiritually informed practices, practices that not only express identity and give meaning, but which also provide a knowledge base that could be helpful in developing sustainable resource management practices. The market-based approach to water management instituted by the Chilean government has been an all-or-nothing proposition, allowing no room for the Rapanui to continue their way of life or contribute to broader conversations about the sustainability of the island's most critical resource.

A colonial legacy of spatial exclusion

While today Rapa Nui is part of a territorial organization that reflects political tensions, overlapping laws, and institutional jurisdictions, the island's original population and its resource base have experienced a variety of mixed management arrangements under Chilean territorial sovereignty. Between 1862 and 1864, prior to its annexation, the island's population suffered a near-extinction at the hands of the Pacific slave raids (AMBAR, 2001). In 1888, Rapa Nui became Chilean. At the time of the island's annexation, a violent segregation was imposed on the Rapanui population as transnational corporate interests turned the island into a livestock ranch. Hanga Roa, the island's only municipality, was walled and the Rapanui population confined within its perimeter (Cooprogetti, 2005; Fischer, 2010). The island functioned as a sheep ranch until 1917, when it was declared a sub-delegation of the Naval Department of Valparaíso, which effectively made Navy officials representatives of the Chilean government (Delsing, 2012). During much of the twentieth century, exclusionary practices by the Chilean Navy's strategic use of the island continued to make the Rapanui people subjects of what has been referred to as *explicit exclusion* (Sibley, 1995).

Today, a less visible kind of spatial exclusion operates and is "taken for granted as part of the routine of daily life" (Sibley, 1995, p. ix), while continuing to suppress vital cultural knowledge and practices. The majority of the island is federal land (*tierra fiscal*), mostly managed by the government as a national park under the CONAF (National Forest Corporation) and protected from all settlement, including by Rapanui. The remaining 30 percent of the federal land is known as Fundo Vaiter and is administered by corporate SASIPA (Sociedad Agrícola y Servicios Isla de Pascua Ltda). A private subsidiary of Chile's development agency CORFO, SASIPA also manages the island's water and energy (Cooprogetti, 2005) and has reportedly received minimal state oversight while tightly controlling access to water quality information (Milad Garrido, 2010). Land access restrictions are inherent in the 1966 *Ley de Pascua*, which recognizes the right to land ownership by residents of Rapanui descent but limits available lands for restitution to the small urban area of Hanga Roa, effectively preventing the majority of the community from accessing ancestral lands.[2]

Weak regulation of practices and activities connected to the tourist industry, also controlled by private investors, furthers these dynamics of exclusion. The island's national park is reserved for tourists. Archaeological sites hold exclusive access for research, and selected locations are set apart for the Chilean military. Figuratively speaking, tourists, archaeologists, and soldiers traverse the spaces today like "new sheep" freely roaming the island (Young, 2012, p. 20), while the "Rapanui are … trammeled by the 'discursive imperialism' of archeology, tourism, and state bureaucracy that is circulating" (Smith as cited in Young, 2012, p. 20). The land tenure system imposed by the Chilean State, which began with its territorial claim of sovereignty, clashed—and continues to do so—with Rapa Nui cartography of kinship, which remains fundamental to the community's sense of collective family ownership. In other words, Chilean *modernization* and *development* resulted in a spatial exclusion that has disrupted an entire way of life based on adaptation to the island's unique ecology, an adaptation that had evolved over centuries.

Exploring the implications of an institutional setting "shaped for the market"

As mentioned in the introduction, Rapa Nui, though geographically isolated, is embedded in an institutional landscape defined by Chilean law and policy; but this institutional landscape also reflects the powerful influences of international lending institutions such as the World Bank. Understanding the relationship between a nationally embraced development paradigm and international financing agencies is central to understanding the implication of Chilean policy for local communities. It is also central to understanding the broader relevance of those communities' experiences, in this case the Rapanui, when it comes to the management paradigms of national governments and international lending institutions, and the principles and best practices they embody. For example, in a 2011 interview about international water policy, a senior water advisor at the World Bank reasserted the widely held belief in the efficiency of water markets and trading water rights in Chile.[3] He coupled this assertion with his own longstanding view of the country as a success story for the privatization of water utilities. He described the Chilean government's process as follows:

> First they corporatized the big water companies. They then put a regulator in place, who for ten years learned to do business in the relatively easy

> environment of regulating state companies. Once the regulator was strong and experienced, they privatized, with great success. I would summarize this as follows. The key underlying determinant is the seriousness and capability of the government.
>
> (Briscoe, 2011, p. 150)

At first glance, the position expressed in this quotation may appear unremarkable, if not self-evident: water is an economic good that needs to be efficiently allocated and managed, a process best carried out by the private sector. The "seriousness" and "capability" of the Chilean government cited by the World Bank advisor refers to the government's commitment to privatization with a minimization of market interference. These management principles theoretically guide decision-making based on the economic efficiency of allocation at varying levels of scarcity. They presume to represent the optimal, if not ideal, guide for managing water regardless of setting and conditions.

However, the government's "seriousness and capability" is a direct reference to practices and policies developed at both the national and international levels during General Augusto Pinochet's two-decade rule of Chile. The Pinochet Administration, which crafted the 1980 Constitution and the 1981 Water Code, operated under strict autocratic conditions, arguably a controversial context.[4] Chile emerged as a neoliberal example of development within a larger global project of neoclassical capitalism, where intensification of a monetarist culture permeated every aspect of daily life as both an economic and a cultural project (Carcamo-Huechante, 2007; Leff, 2010).[5] During this period, Chile modeled its policy on market economics, which the World Bank viewed as a "great achievement" and "a bedrock on which to refine Chilean water management practices" (Briscoe, Anguita-Salas, & Peña, 1998, p. 9). Notwithstanding Chile's extraordinary political context, the economic establishment treated its policy approach as a solution to the universal problem of effectively managing water resources as an economic good—a solution other countries would ideally adopt.

Even if one considers it possible to view Chile's policies in isolation from its military government, the Chilean model of water markets has clearly produced conflicting outcomes. For example, Chile has succeeded in meeting the Millennium Development Goal Target for water coverage and sanitation, as defined in the 2015 report, having achieved some of the highest levels for those parameters in Latin America (WHO/UNICEF, 2015). But nationally aggregated numbers blur the realities of local communities rooted in diverse ecological and cultural settings. Most think of this nationally aggregated outcome as the result of an economic model that has championed free-market reform of law, economics, and policy to the extent that institutions have been "shaped by and for water markets" (Bauer, 2010, p. 45). However, contrary to expectations, access to water has become a major source of conflict across the country and has moved to the forefront of national policy debates around problems ranging from ecosystem damage to inequitable access and allocation patterns (Bauer, 2014; Larraín & Poo, 2010; Redagrícola, 2014).

In places like Rapa Nui, where citizens have been excluded from the exercise of many of their civil, cultural, and/or social rights, regimes that theoretically assume full exercise of those rights require serious scrutiny. It can be instructive to contrast the results of institutionalizing market-driven policies under the aegis of authoritarian power with the experience of nations that define water as a common good in the European tradition of universal access to health, education, safe water, and sanitation, where multiple values of efficiency, access, and rights are served (Castro, 2006).

The experience of the Rapanui embodies a larger Latin American paradox. While this region of abundant hydrological resources represents 33 percent of the world's water capital (World Water Council, 2015), some 50 million people lack adequate water services, and 110 million of the region's population lacked acceptable sanitation in 2012 (UNICEF, 2014).[6] Troubling as these discrepancies are, they are but the tip of the proverbial iceberg regarding inadequacies that extend beyond the technical problems of resource management.

These deficiencies also manifest in a broader lack of social and civil services—the material emblems of citizenship—in turn, limiting the scope of citizenship rights. At the same time, lack of access to clean water and sanitation in the twenty-first century signals the political invisibility of so many people in the region (according to Natanson's interview of O'Donnell, 2006). As such, water inequality is inseparable from the deeply embedded cultural and social problems of exclusion and political invisibility that undermine claims of development models. In Chile, these models have centered on economic growth while treating contextual social, political, ecological, and cultural factors as externalities.

Valuing water in context: towards a plurality of solutions

We now return to our earlier question of how valuing water as an economic good responds to both water's complex nature and the legitimacy of diverse allocation patterns regarding water rights. When viewed as a commodity, water would appear to be well suited to principles of management, principles concerned with scarcity as well as with efficiency of allocation, use, operation and maintenance, and, when necessary, the development of new sources. But water is also increasingly recognized as a complex resource integral to the larger earth system and necessary to sustain life itself, requiring an approach that includes management of ecosystem health, partnership, and good governance (Schultz, 2014).

Addressing water management through good governance and partnership represents a paradigm shift away from capital-intensive, private sector investments in global water supply evaluated on financial return, with most decisions taking place behind closed doors, and a shift towards participatory processes driven by a variety of considerations rather than only by technical factors (Aguilera-Klink, 2008; Conca, 2006). Water physical scarcity has often been the exclusive concern addressed by technological solutions, ignoring circumstantial shifts in the larger system due to social or cultural contexts. In

Chile, the assumption that private sector management is the most efficient and rigorous way to address these issues runs deep.

Structural changes anchored in the 1980 Chilean Constitution are reflected in the state's economic and institutional organization, which have contributed to social transformations of water as a complex system. These social transformations are relevant to assessing water market effectiveness in managing water's complexities. Indeed, within the framework of water market efficiency, critical contextual shifts have gone ignored (Bauer, 2014). For example, Chile's Constitution has yet to recognize indigenous rights, and without those rights, the water system suffers from a critical deficiency that distorts notions of effectiveness and efficiency in access and distribution. When, in 1988, a national plebiscite brought an end to the 17-year regime of Augusto Pinochet, a transition to democracy was set in motion. Over the next decade, Chileans experienced an increasing number of economic measures aimed at advancing free trade and privatization strategies aligned with the neoclassical program of development initiated by the Pinochet regime. These policies enabled a universal approach to resource management, one that has also caused increased environmental degradation and the dissolution of traditional practices (Leff, 2010).

Chile's "transition to democracy" also saw the advancement of an indigenous and human rights movement—*derechos de los pueblos originarios*—which gave rise to a formal recognition of Chile's multiethnic nature.[7] Despite progress in certain areas of civic life, such as education and self-representation, tension rose as different goals came into conflict over resource appropriation, land tenure, and the reassertion of collective identities. In these and other conflicts, indigenous communities' claims did not conform to the neoclassical program. Turning to government institutions to advance and protect their community integrity, indigenous communities, including the Rapanui, found themselves appealing to a state in which the interests of private capital and foreign investment dominated. Despite formal legislative recognition of territorial, social, economic, and cultural rights for indigenous people, these rights became subordinate to the primacy of privatization and growth.[8] As a result, ongoing efforts in Rapa Nui to protect these rights generally failed to achieve legally binding action on behalf of islanders, thereby weakening the ethno-political dimension of the debate and undermining the rights-based policies established under the 1993 Ley Indígena, including the right to fresh water.

Earlier, we also asked how generalized Chilean policies and practices regarding water scarcity and management efficiency impact the life of the Rapa Nui community today. As mentioned, the island's hydrological resources are at risk (AMBAR, 2001; Calderón-Azorar & O'Ryan, 2011; Figueroa & Rotarou, 2013). Fresh water and drinking water originate from deep wells that, while providing a plentiful volume/resident ratio, increasingly supply water of questionable quality, thereby shifting the focus on volume of the aquifer's carrying capacity to one of quality risk. Available technical literature along with recently published field observations make clear that, because of its vulnerability, Rapa Nui is unsuitable for sustaining current levels of tourist activity—the island's main source of revenue.[9]

Rapa Nui's permanent population of roughly 5,600 people has seen steady but average growth, doubling in size in twenty years (AMBAR, 2007), but the tourist population and transient personnel from continental Chile continues to grow by the thousands. For example, CONAF official reports indicate park attendance growth from 17,000 in 2002 to 50,000 in 2011, roughly a 300 percent increase over the course of only a decade (Woolson, 2014). Other reports, which point to a lack of planning in tourist industry development, place tourist numbers at 100,000 annually (Calderón-Azorar & O'Ryan, 2011). In interviews, park rangers reported over 70,000 tourists in 2012, and community members have expressed alarm that "tourism is out of control" (as cited in Woolson, 2014, p. 197). Growth in tourism has led to increased sources of pollution and pressures on the land that are leading to ecological change, and the growth is occurring with no clear vision or accompanying regulation to temper its impact and sustain the island's carrying capacity.

Growth in tourism has intensified point source pollution, a serious problem directly contributing to the degradation of the island's water resources. Sources of water pollution include a landfill lacking protective membranes for municipal waste disposal; no sewer system and increased number of cesspools; lack of collection sites for fuel; contaminant infiltration from quarry use of explosives; changes in land use; over-extraction leading to salt water intrusion; and a rapid increase in plastics recyclables that cannot be returned to the continent due to health restrictions (CONAF as cited in Woolson, 2014, p. 198; ECOPOLIS, 2010; Milad Garrido, 2010). This level of system complexity is beyond the reach of efficient management.

Meanwhile, residents report poor water quality and lack of access to information about the water they consume (Figueroa & Rotarou, 2013). SASIPA's official reports on quality control show no contamination, but testing primarily measures salinity levels rather than other contaminants (Montgomery, 2011). In interviews, I recorded residents' low confidence in the company's practices and their concerns about existing social tension, particularly between residents and state-sponsored private control (Woolson, 2014). Park administrator E. Tucki explained that "there is a lack of a shared vision for the island, a lack of governance" (Calderón-Azorar & O'Ryan, 2011, p. 7). The disproportionate growth of tourism reflects the dominant political landscape, and in particular policy development that enables business growth without meaningful public participation or transparency. Instead, the community calls for "a joint project that should have civil society actively participating" in decision-making (museum director F. Torres as cited in Calderón-Azorar & O'Ryan, 2011, p. 7). Such a collaboration would achieve what is known as *legal pluralism*, or "the art of seeing the same thing from two different perspectives," when developing policy (Aldea Educativa Principal as cited in Woolson, 2014, p. B8). These comments lead back to a central theme in this analysis: how context and complexity are inherently connected to concepts of citizenship and democracy, as they represent dynamic and adaptive aspects of water governance essential for rethinking management.

Concluding thoughts on governing complexity for a plural society in a global landscape

As we have seen, unconstrained expansion of Rapa Nui's tourist sector stands at odds with the integrity of its critical ecosystem and the clean water that sustains life on the island. But growing ecological impact to the aquifer also undermines the environmental and cultural rights of the community, thereby producing a kind of socio-spiritual effect that is in conflict with sustainability. At the same time, pressures are multiscalar: while Chilean state policies have led to material appropriation of the territory and its resource base, including water, the island's international World Heritage designation has resulted in a conceptual appropriation of assets and identity that has, in turn, fostered the global imagination of Rapa Nui as a "museum island" (Fischer, 2010, p. 199).

Within this troubling context, land access and land use remain a point of contention between the community and the state (Cooprogetti, 2005; Delsing, 2012; Young, 2011). Unfortunately, these conflicts are contingent on a process that reproduces the Rapanui as colonial subjects and fuels a gradual process of acculturation. Pinochet-sponsored laws constituted a clear effort to replace collective land ownership by redefining Rapanui households within a framework of privatized individual rights. Successive laws and decrees emphasizing privatization include the administration and distribution of water, making access to clean, fresh water subject to intensified debates (Calderón-Azorar & O'Ryan, 2011; Woolson, 2014). This shift toward private ownership has adversely affected the Rapanui community, weakening the longstanding cultural resilience displayed for more than a century, while increasing differences among local leaders. As Delsing has articulated, "The separation of Rapanui from their ancestral lands (*kainga*) and the relentless insistence on the benefits of private landownership has eroded Rapanui customary law" (2012, p.73). My own interviews reveal not only a fractioning of Rapanui society, but also an intensification of that fractioning around material possessions and political privilege.

Understanding water as a complex system requires that we incorporate context—in this case, the Rapanui context. Studying the island's dynamic carrying capacity as it relates to tourism demands is not enough. We must also look at the impacts of pollution and cultural loss, including loss of the Rapanui knowledge system, which does not conform to the symbolism of Western tradition. On the island, two cosmologies, or ways of viewing the world, coexist: one Eurocentric, or "continental", and the other non-European in origin. As mentioned earlier, water is constitutive of both a natural cycle and a spiritual life cycle. For more than three centuries, the Rapanui cosmology has supported adaptive practices to the arid environment, reflecting a wealth of traditional knowledge inherent to a culture of modest consumptive habits built on generations of adaptive living. As one informant explained:

> The Rapanui are well trained in living. They know all the resources that can help with daily life. They adapt well. When the sea is good, I am a

> fisherman. If it gets bad and it rains, I am craftsman. And if the rain stops but the sea is bad, then I am a farmer. Well trained in living.
>
> (Informant 4 as cited in Woolson, 2014, p. 208)

Or one could assert, trained in sustainable living.

In terms of citizenship rights, Rapanui today are finding ways to assert their identity and negotiate their rights to self-determination. They are valuing their holistic relation to the material world of nature. For example, the national park and its staff support the work of Rapanui tour guides, whose role is self-described as "hosts" or "teachers" of a cultural patrimony that has always welcomed tourists as "guests to our home" (informant 5 as cited in Woolson 2014, p. 204). In addition, they promote the recovery of native plants for reforestation, resisting use of continental species prescribed from mainland offices, and they encourage the use of stone gardens.

Stone or rock gardens are an example of human adaptation to a dry volcanic environment, and Rapanui have been cultivating these for centuries (Stevenson *et al.*, 2006). These gardens, or *Manavai*, make use of rocks for moisture control, temperature regulation, and soil and plant protection—and they require no irrigation. Minerals in the stones supply nutrient to the plants that, while smaller in stature than average, boast a higher concentration of nutrients and calories. This sophisticated knowledge, in which the *rock* represents survival, included seven types of garden structures confirmed by the archaeological record (Haoa & Gonzalez, 2010). Archaeologist Sonia Haoa explains how these traditional and sustainable practices contrast with conventional agriculture with its dependence on irrigation and land clearing:

> What took men hundreds of years to do, we remove in minutes, causing flooding and erosion. Our soil has lost a meter of depth because we have totally reversed the concept [of survival]. Understanding how humans adapted to this environment is beautiful and a luxury that should be our lesson: an adaptation of human beings to the environment in total isolation.
>
> (p. 8)

To rethink management in contextually productive ways means considering a comprehensive governance strategy involving all actors—the government and the community—and embracing democratic practices as dynamic aspects of the decision-making process. Management tends to seek efficiency relative to scarcity. And scarcity is, in many ways, a twenty-first-century development for the Rapanui, who have lived on their land with a sense of abundance. Indeed, for more than three centuries, the lack of rivers and surface water did not prevent life on the island from flourishing (Roggeveen as cited in Fischer, 2010, p. 27 (Original Logs of 1721–22, published London, 1908)).[10] More importantly, valuing water exclusively as an economic good does not respond well to the complex nature of water in a plural society with dynamic ecosystems; a technical approach to water may evaluate it, quantify it, and qualify it, and, as a result, codify it to be managed in a particular way, but water undergoes both physical and social transformations that require a complex understanding of its particular settings.

In conclusion, Rapa Nui is an observable "microcosm" that represents a manageable example of thinking systemically about water, an inherently complex and multiscalar resource. Governance is therefore best suited for addressing the many layers of Rapa Nui water as described in this chapter, including its spiritual designation as part of life itself. Recognizing how different groups relate to water also calls for engaging inclusive ethical principles that respect diversity. Of course when it comes to water, the ethics of diversity is not a new idea: a water scholar notes, "Calls for new water ethics are not new" (Groenfeldt, 2013, p. 155). However, what *is* novel in the twenty-first century is the integration of an emerging water ethic within global governing standards, a water ethic that legitimizes governance roles for everyone within a given water basin (Groenfeldt, 2013, p. 164).

Water governance is then to be understood as the central challenge of embracing water's inherent complexity, not only in theory but also in effective participatory ways. Water governance provides the most adaptive process for addressing water quality, quantity, and access, and it subsumes the more narrow and restrictive paradigm of *water management* in order to facilitate the integration of diverse political and cultural perspectives. This discussion of water in the context of Rapa Nui enhances our understanding not only of the Rapanui people and place, but of water systems more generally, showing that seemingly disparate concepts of cosmology, complexity, and cultural practice meaningfully intersect in the lived experience of people and place and that these intersections support new ways of thinking about water, or perhaps new ways of thinking about a *water culture*, for sustainable living. The water culture that emerges from these new ways of thinking governs the ecosystem; an approach that understands economic activities within the limits of the ecosystem's capacities and integrates historic contexts into social premises for formulating collective solutions (Aguilera-Klink, 2008).

Notes

1 Human right to water is understood as the natural entitlement to access a vital resource for life (UN Human Rights Council Resolution 16/2, 2011). See Glick (1999).

2 Ley de Pascua is the 1966 law that created the Department of Easter Island and brought navy rule to an end. The law designated Hanga Roa as the municipality and recognized Rapanui as Chilean citizens, for the first time (Ley-16441, n.d.).

3 Water markets indicates trading water resources, often interpreted as water rights, where value is produced through pricing (Bauer, 2010).

4 Pinochet became Chile's established ruling general on September 11, 1973, through a violent coup.

5 Neoliberalism is understood as a capitalist model of flexible accumulation characterized by new information technologies, emphasis on the consumer, service workers, feminization of the workforce, and globalization of financial markets.

6 Statistical numbers vary. The 2015 forum indicates that only 20 percent of wastewater is effectively treated.

7 Law N° 19.253, or Ley Indígena (1993), recognized Chilean indigenous people as "human groups that have existed in national territory since pre-Colombian times

and that have preserved their own forms of ethnic and cultural expression, the land being the principal foundation of their existence and culture" (UN, 2003, p. 2). Other institutions that followed include the National Corporation of Indigenous Development (CONADI), Commission of Historic Truth and New Relationships with Indigenous People (2001), ratification of the 1988 "Agreement of Wills," and the Commission for Development of Easter Island (Delsing, 2012).

8 See Delsing (2012) and Young (2012). The 1979 Decree 2536—Chile's "anti-communal indigenous land ownership" decree—aimed to dissolve communal land in order to advance private investment. Ley Pinochet, or Decree 2885, assigns state jurisdiction over fiscal land and individual property rights to urban land in contradiction to collective ownership.

9 For a review of tourist industry reports, see AMBAR (2001), CONAF (2012), ECOPOLIS (2010), Hydrological reports (DGA, 2011).

10 Historians estimate that the original island settlers arrived in ad 800 and adaptation followed the twelfth century loss of palm tree forests and higher humidity that had characterized earlier centuries (Stevenson, 2006). See also Fischer (2010).

Works cited

Aguilera-Klink, F. (2008). *La nueva economía del agua.* Madrid: Fundación Argentaria.

AMBAR. (2001). *Estrategias y acciones para la conservación, uso y aprovechamiento sustentable de los recursos patrimoniales de Isla de Pascua.* Santiago: CORFO- MIDEPLAN-CONADI.

AMBAR. (2007). Declaración de Impacto Ambiental. Plan regulador comunal de Isla de Pascua. Municipalidad de Isla de Pascua- Ministerio de Vivienda y Urbanismo V Región d Valparaíso.

Bauer, C. (2010). Market approaches to water allocation: Lessons from Latin America. *Universities Council on Water Resources Journal on Contemporary Water Research & Education, 144*, 44–99.

Bauer, C. (2014). ¿La ley del péndulo? Conflictos de agua y gobernancia en Chile desde 2005. In Montenegro et al. (Eds.), *Actas de las VII Jornadas de derecho ambiental. Recursos naturales ¿Sustentabilidad o sobreexplotación?* (pp. 637–673). Santiago de Chile: Legal Publishing, Thomson Reuters.

Bauer, C., & Prieto, M. (2012). Hydroelectric power generation in Chile: An institutional critique of the neutrality of market mechanisms. *Water International, 37*, 2, 131–146.

Briscoe, J. (2011). Two decades at the center of world water policy: Interview with John Briscoe by the Water Policy Editor-in-Chief. *Water Policy, 13*, 147–160.

Briscoe, J., Anguita-Salas, P., & Peña, H. (1998). Managing water as an economic resource: Reflections on the Chilean experience. Environment Department Papers, 62. Washington, DC: World Bank.

Calderón-Azorar, C., & O'Ryan, P. (2011). Desafíos y oportunidades de desarrollo sostenible de Isla de Pascua basadas en turismo. RedSur Consultores, Santiago de Chile/Hanga Roa.

Cárcamo-Huechante, L. (2007). *Tramas del Mercado: imaginación económica, cultura pública y literatura en el Chile de fines del siglo veinte.* Santiago, Chile: Editorial Cuarto Propio.

astro, J. E. (2006). *Water, power, and citizenship. Social struggle in the Basin of Mexico.* New York: Palgrave Macmillan.

Corporación Nacional Forestal. (2012). *Estadísticas de visitantes años 2004–2011* (CONAF 2012 National Parks Reports). Santiago, Chile.

Conca, K. (2006). *Governing water.* Cambridge, MA: The MIT Press.

Cooprogetti, S. R. L. (2005). *Informe preliminar Hotel Explora Isla de Pascua. Estudio de Impacto Ambiental.* Buenos Aires: Cooprogetti Argentina.

Delaune, G. (2012). Rapa Nui on the verge: Easter Island's struggles with integration and globalization in the information age. *Berkeley Planning Journal, 1*, 25, 126–139.

Delsing, R. (2012). Issues of land and sovereignty: The uneasy relationship between Chile and Rapa Nui. In F. Mallon (Ed.), *Decolonializing native histories* (pp 54–78). Durham, NC: Duke University Press.

DGA. (2011). Minuta plan especial Isla de Pascua. Santiago. DIRPLAN, Departamento de Planificación. Retrieved from http://documentos.dga.cl/SUB5237.pdf

ECOPOLIS. (2010). Hacia una Rapa Nui integrada y sustentable. Hanga Roa, Isla de Pascua. Ecopolis.

Figueroa, E., &. Rotarou, E. (2013). Environmental impacts and challenges of tourism development on Easter Island. *Gran Tour,* 7, 39–59.

Fischer, S. (2010). *Island at the end of the world.* London: Reaction Books.

Glick, P. (1999). The human right to water. *Water Policy, 1*, 5, 487–503.

Groenfeldt, D. (2013). *Water ethics. A values approach to solving the water crisis*. London and New York: Routledge.

Haoa, S., & Gonzalez, L. (2010). Los jardines de piedra. *Moe Varua. Rapa Nui, 1*, 6–9.

Haun, B. (2008). *Inventing "Easter Island."* Toronto: University of Toronto Press.

Hunt, T., & Lipo, C. (2012). Ecological catastrophe and collapse: The myth of "ecocide" on Rapa Nui (Easter Island). In P. McAnany & N. Yoffee (Eds.), *Questioning Collapse. Human resilience, ecological vulnerability, and the aftermath of empire* (pp. 21–44). New York: Cambridge University Press.

Larraín, S., & Poo, P. (2010). Conflictos por el agua en Chile. Retrieved from http://cl.boell.org/sites/default/files/05_conflictos_por_el_agua_chile.pdf

Leff, E. (2010). *Discursos sustentables.* México: Siglo XXI Editores.

Ley-16441 (1984). 01-mar-1966 Ministerio del Interior – Ley Chile – Biblioteca del Congreso Nacional. Retrieved from https://www.leychile.cl/Navegar?idNorma=28472

Milad Garrido, C. (2010) Diagnóstico de los recursos hídricos subterráneos de Isla de Pascua. (Doctoral Dissertation) Retrieved from Universidad de Chile.

Montgomery, E. (2011). Condiciones hidrogeológicas Isla de Pascua Chile. Gobierno de Chile, DGA. Retrieved from http://documentos.dga.cl/SUB5237.pdf

O'Donnell, G. (2006). Sobre los tipos y calidades de democracia. Entrevista a Guillermo O'Donnell/ Interviewer: J. Natanson [Transcript], Página 12. Retrieved from http://www.pagina12.com.ar/diario/dialogos/21-63632-2006-02-27.html #

Redagrícola (2014). Desde aguas abajo: La batalla de agua de Copiapó. Retrieved from http://www.redagricola.com/reportajes/agua/la-batalla-del-agua-de-copiapo

Rivera, C. (2012, June 21). Tres claves para entender por qué Isla de Pascua está al borde de un colapso medioambiental. *El Dínamo*. Retrieved from http://www.eldinamo.cl/pais/2012/06/21/3-claves-para-entender-por-que-isla-de-pascua-esta-al-borde-de-un-colapso-medioambiental/

Schultz van Haegen, M. (2014). The Dutch minister of infrastructure and the environment. Speech presented at UNSGAB Meeting, 02-06-2014, Singapore.

Sibley, D. (1995). *Geographies of exclusion*. London: Routledge.

Stevenson, C., et al. (2006). Prehistoric and early historic agriculture at Maunga Orito, Easter Island (Rapa Nui), Chile. *Antiquity, 80*, 919–936.

UNICEF. (2014). Progress on drinking water and sanitation. Retrieved from http://data.unicef.org/resources/progress-on-drinking-water-and-sanitation-2014-report.html

United Nations Economic and Social Council. (2004). Executive summary of the Report of the Special Rapporteur on the situation of human rights and fundamental freedoms of indigenous people. Resolution 2003/56, Addendum. Retrieved from http://acnudh.org/wp-content/uploads/2011/01/Mission-to-Chile-2003.pdf

WHO/UNICEF. (2015). Joint monitoring programme for water Supply and Sanitation. Retrieved from http://www.wssinfo.org/data-estimates/tables/

Woolson, M. A. (2014). El espacio como espejo cultura (doctoral dissertation). Retrieved from Department of Spanish & Portuguese, University of Arizona, Tucson, AZ.

World Water Council. (2015). Seventh world water forum, Korea 2015. Retrieved from http://www.worldwaterforum4.org.mx/home/main_results.asp?lan

Young, F. W. (2011). Unwriting Easter Island (doctoral dissertation). Retrieved from Department of Anthropology, University of Hawaii, Honolulu, Hawaii.

Young, F. W. (2012) 'IHe Koe? Placing rapa Nui. *The Contemporary Pacific*. 24:I

15 Strange waters

From confluence to vortex in the Los Angeles basin

T. S. McMillin

The day's fieldwork begins at the Los Angeles River Center and Gardens, where I park the red rental car and run into Lewis MacAdams. The co-founder of Friends of the Los Angeles River (FoLAR), MacAdams is, to my mind, *the* friend of the LA River, a poet and scholar who, for the past several decades, has been a principal activist on behalf of (and self-proclaimed spokesman for) the stream I am researching. I introduce myself, although not because I'm outgoing: the cookie from last night's restaurant contained a line from Ralph Waldo Emerson, "The only way to have a friend is to be one," and since I am here to understand the river, I've taken my good fortune as a nudge to befriend one of the river's best friends. MacAdams kindly agrees to meet with me later in the week, and before I set off on foot to discover the river for myself, he warns me to drive farther upriver to a more scenic section. Around here, he says, it looks like a construction zone. But I've always wondered how rivers were made, so on foot I go—around the corner and down Figueroa, climbing over concrete barriers, edging along orange plastic fencing, squeezing through a narrow passage between the railing and fleetly passing cars—to the middle of the Riverside Bridge looking out on the midstream building project near the river's confluence with the Arroyo Seco.

The confluence serves as a centerpoint for my experiment regarding the nature of the Los Angeles River. Twofold in its approach, the experiment combines fieldwork on the current conditions of an urban waterway with research on textual representations of rivers. The primary goal of the experiment is to investigate the ways in which humanistic inquiry might serve as a means of conceptually "daylighting" a river. "Daylighting" most commonly refers to the process of opening or redirecting the channel of a stream that has been covered or obstructed by human structures, a physical return to the light of day. Inspired by the term's metaphorical potential, I have adopted it as a key component in an area of study I call "textual potamology": an exploration of the river and its history in an effort to imagine new ways of understanding a river's nature. Textual potamology pursues the possibility of bringing to light not only the river but our thought processes, our social practices, and our cultural biases regarding waterways. In this essay, I will consider some ways in which nonfiction and its interpretation might cast some daylight on the LA River.

On the off-ramps and on-ramps and surface streets, walking where one really ought not to walk, I am surrounded by the roaring engines and compounded energy of the freeways. Though the morning rush hour has abated, the tumult remains constant: freight trains rumble, construction grinds on, an Amtrak screams. Layers of incessant movement, tiers of concrete, lines of power. Power manifests multifariously: that which drives the cars, that which formed the structures that channel the cars, that which fixed the course of the river and cemented the confluence into place. The Los Angeles River, rebuilt into the Los Angeles Flood Control Channel and maintained by such entities as the Los Angeles Department of Water and Power and the U.S. Army Corps of Engineers, is an apt emblem for all of LA. This place is an enormous system of engineered cuts and grooves, sluices and outlets through which courses "power"—the ability of humans to alter the physical aspects of a place. The story of Los Angeles would appear to be a tale of "culture" controlling "nature," of culture built up, layer by layer, on top of nature. I am investigating the river in order to test the veracity of that story.

On a map, the Los Angeles River looks something like an upside-down, reversed, and squiggly L. From its headwaters in the west end of San Fernando Valley, the river runs to Glendale and then bends sharply to the south. The center third, where my current fieldwork is based, encompasses the traditional site of the city's founding and includes sections called Glendale Narrows and

Figure 15.1 Confluence, looking upriver

Source: Author.

Elysian Valley, as well as the river's confluence with the Arroyo Seco (a Spanish name meaning "dry wash"), where I am now standing. Not far from here, Father Juan Crespí wrote in his field notebook on August 2, 1769: "Another good-sized, full-flowing river with very good water, pure and fresh, flowing through another very pleasant green valley lying westward" (Crespí, 1771/2001, p. 337). Crespí, looking back on the places he had already visited on La Sonoreña, his red mule (p. 52), thinking of what lay ahead, but most of all gazing fondly at the lands and people around him, noted, "We have proclaimed it The River and Valley of *Nuestra Señora de los Ángeles de la Porciúncula*" (p. 339)[1] and declared it a "very lush pleasing spot, in every respect ... a most beautiful garden" (Crespí, 1771/2001, p. 337). He was dizzied with delight, relieved that the women of these parts were more decently clad than previous "heathens" (p. 339), and only a trifle concerned with the fourteen "very persistent and strong though not long-lasting" quakes they had experienced during their short time in the area (p. 343).

The *padre* and his party were drawn to this place by the waters of the *rió* he took to calling *La Porciúncula*. The "very tractable, friendly Indians" (p. 341) with whom his party met were descendants of the Tongva, or Kizh, another people who had been attracted by these waters over a thousand years earlier, following Hokan-speaking peoples who had been drawn there 10,000 years ago. I too have come to Los Angeles for the river, though less for its "pure and fresh" (p. 337) contents than for the strangeness of its waters.

Running fifty-one miles from a concrete confluence of tributaries in Canoga Park and emptying into the Pacific at its engineered mouth in Long Beach, the LA River has undergone diversion of its flow, widespread clearing of the flood plain, the proliferation of surface-paving, and the depletion of underground sources. This led to bringing water to Los Angeles from sources far away, which contributed to a sense that the existing river was no longer much of a river. As its age-old practice of periodic flooding began to have an increasingly catastrophic effect on the city's burgeoning population, the massive flood-control projects of the twentieth century made it seem even less like a river, furthering its neglect by residents of and visitors to the greater metropolitan region. Recently, however, activists such as MacAdams and local officials have set about politically—and in some cases physically—*re-rivering* Los Angeles. In 2010, the Environmental Protection Agency (EPA) declared the river's waters officially navigable, overturning prior declarations to the contrary by the U.S. Army Corps of Engineers and other governmental entities, thus granting the river and its tributaries protection under the Clean Water Act. In 2013, a section of the river not far from here was officially declared a recreational zone—suitable for boating and fishing but not swimming—due in part to the success of a program begun in 2011, when authorities began issuing permits for weekly canoe/kayak trips on certain sections upriver. I was invited to serve as an educator on two such float-trips with LA River Expeditions ("Putting the river back in the river"; lariverexpeditions.org) near Sepulveda Basin in 2012 and can testify that, in many ways, the LA River offers a case study of the "life, death, and possible rebirth" (Gumprecht, 1999) of urban streams. Still, whether dead or alive or somewhere in between, the Porciúncula poses questions.

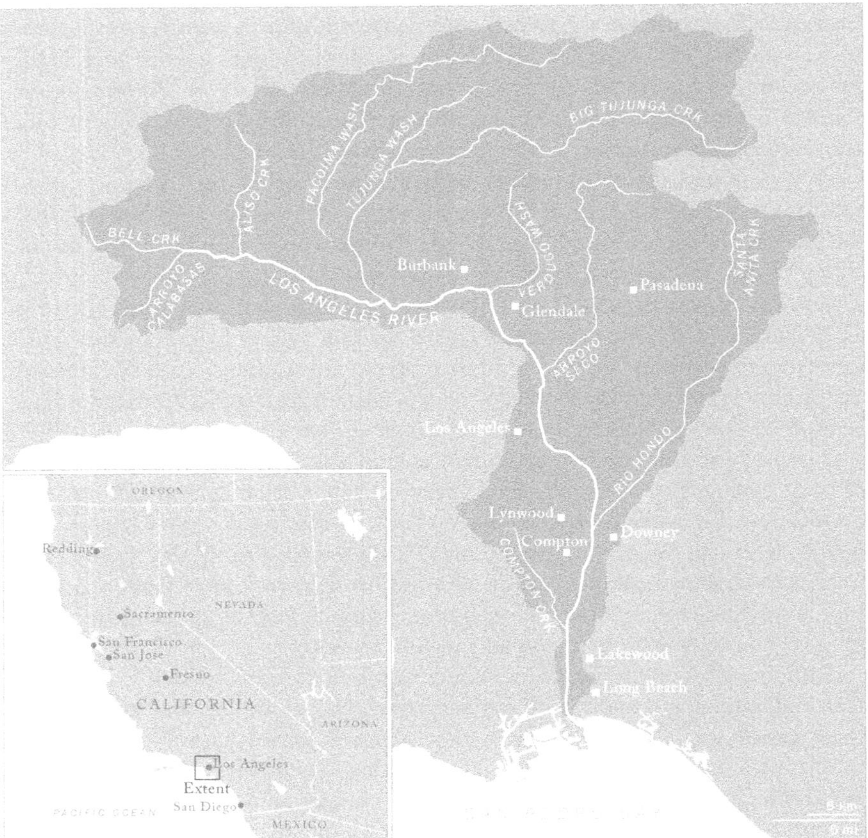

Figure 15.2 Map of the Los Angeles River Basin

Source: Cartography by Gregory Woolston using data from (Shannon1, 2011), Natural Earth, and © OpenStreetMap Contributors.

Standing at the confluence of river and arroyo beneath the traffic-trembling bridges of the 110, America's first freeway, I feel as if centrifugally pulled to the historical and geographical center of the river in the heart of the most heavily populated county in the United States. On the riverbed at midday when the equipment rests, however, the thin flow of the arroyo runs into the altered course of the main stream in a clandestine meeting well beneath all the action. Under LA, in the shadows of various structures, the confluence seems a separate world. Looking downriver towards Chinatown, the bed appears empty to me, an untraveled road crossed by multiple bridges, as if river and arroyo had converged into nothing. Train after train roars along the west bank, a contrast that renders the emptiness emptier. Beside the confluence, Arroyo Seco trash collects: caps, tennis shoes, golf balls—reminders of events up creek in Pasadena. A fire smolders on the north bank, a remnant of life in the half-dozen homeless shelters under the bridges; dogs—quiet, watchful as I poke about—sit by tents, while an older man arranges his belongings. Across the arroyo, the construction equipment is laid out straight and ready. Down the center of the tributary's channel, an inch or two of a trickle slides along the concrete; looking up its course, I see a younger man brushing his teeth. He is gone by the time I make my way up there, but his laundry hangs nearby. From a bridge spanning the creek, I can look up into the hills, see other bridges crossing close to the top of the banks and heavy foliage covering the margin of the thin ridges, but the bed of the arroyo itself is blank, a stark stream of nothing. I can see graffiti on the distant spans but can't make out the words.

★★★

From the stillness of the arroyo bed, I can sense some of the forces that make the confluence an "American narrative vortex," in the words of Jenny Price. Her essay "Thirteen Ways of Seeing Nature in L.A."[2] has its own sort of power. It brought the basin's strange nature to my attention, serving as a conceptual map of the river's course and its place in the region, as an introduction to the social history of the river, and as an intellectual guidebook to the place and some of its problems. Price's essay exemplifies some of the diversity and significance of non-fictional literature. From Crespí's eighteenth-century description of the river to Price's twenty-first-century urban nature writing, non-fictional prose has brought together people, place, stories, and ideas in a momentous swirl.

Unfortunately but perhaps not surprisingly, a great deal of nonfiction regarding LA has omitted discussion of the river, while many a work that does mention the river "imagines nature in opposition to cities" (Price, 2006, vol. 2, p. 32). Nonfiction—fact-based writing, a writing that is not fabricated but situated in reality—encompasses myriad forms, dealing with real people (memoir, biography), real events (journalism, history), and real places (guidebooks, travel accounts). Considered broadly, the category extends from maps and manuals to studies and essays (including those in this volume). Functions differ widely within this category, but nonfiction invariably engages the world as we know it and is interested in the truth: capturing, cataloging, collecting, documenting, investigating, charting, or reporting it. Nonfiction *corresponds* with the real world. It indicates, notes, underlines, or explains, drawing attention to specific aspects of a particular topic or object of focus. But also, in some forms, it connects parts into a meaningful whole, not only analyzing but synthesizing as well. Nonfiction writing *maps* the world we think we know. Like the engineered river channel, nonfiction about LA can also be seen as an exercise in power, colorful arrangements of lines that represent a culture's attempt to "get" nature: to locate it, situate it, hold it still for a moment. Nonfiction can be a net for capturing nature, to be used for gathering information, for further study or planning, for guidance, inspiration, or control.

Price's "Thirteen ways" attempts to accomplish some of the traditional goals of nonfiction but with important differences. The essay 1) directly questions the nature–culture opposition; 2) draws attention to the inability of writing—

even nonfiction—to operate as a transparent medium through which the reader can be given the pure facts or complete truth about a place; and 3) explores the ways in which nonfiction might function as a conceptual mapping to help us find "hidden" or "lost" nature. Her argument, in part, is that nonfiction about nature in the United States has tended to overvalue a particular notion of the "wild." "Nature," in much of that writing, is opposed to "culture," becoming something other than human, separate from humans, to be utilized by, exploited by, preserved by, or ignored by humans. A study of non-fictional works about Los Angeles—maps, travel guides, histories—reveals that, through the nineteenth century, the natural attributes of the place made it something of a paradise, a place where culture could thrive. But beginning in the first half of the twentieth century, culture overtook nature in nonfiction about the City of Angels. Emphasis shifted from a Land of Sunshine with a restorative climate to a propitious site for industrial progress and cultural prospects. As a result of this shift, the river faded and eventually disappeared from maps as well as from the pages of guides and histories.

Even as the river has begun to reappear in many non-fictional accounts of Los Angeles, the nature–culture opposition still poses a problem. Price's interest in the physical waters of the Arroyo Seco's confluence with the LA River, a peculiar spot that she herself found only after considerable difficulty, is connected to another conceptual confluence, the viewpoint where culture could (or should) flow into nature. In "Thirteen Ways," after observing that "the stories we tell about nature are the most basic stories we can tell" (Price, 2006, vol. 2, p. 33), she concludes the section with a call for a different kind of story, a different type of imaginative nonfiction: "What we need in L.A., as elsewhere, is a foundational literature that imagines nature not as opposite of the city but as the basic stuff of modern everyday life." This writing would produce "tales that explore our daily, intertwined connections to nature and to each other" (Price, 2006, vol. 2, p. 34). At the Confluence—capitalized in Price's essay—many things come together and entangle: history and present, politics and language, arroyo and river, nature and culture. In pursuit of a better understanding of urban places, "Thirteen Ways" makes a case for using the LA River to think differently about LA's nature, for using nonfiction to foster that thinking, and for the confluence as a point of departure. Informed and inspired by Price's approach, I have begun to see the Los Angeles River as a confluence of confluences, and to think of that mega- or meta-confluence as a kind of vortex, an active, flowing, spinning, growing, changing collection of matter and energy and ideas.

A field of towering rebar vibrates when a Metrolink train rumbles through, and a gap in the shaking vertical lines gives me a glimpse of one of the blue and white Los Angeles River roadside signs with its silhouette of an iconic heron. To get an overview of the confluence, I walk against the freeway traffic, toward the sign and along the river to a spiral stair, up to another level of freeway and across that, and then scramble up the hills through which the 110 cuts its way into LA from Pasadena. I climb a steep bluff to a plateau looking down on the exact confluence: a large rectangular hole in a vertical wall—two such holes side by side, actually, including what appears to be an access point adjacent to the Arroyo Seco mouth, from which trucks and cranes and other heavy equipment emerge and head upriver into the construction zone with its pits and plates and piles and pipes. Below this point, a little way downstream and also a little way upstream, the vertical concrete banks of the river begin to slant, resembling the graphic embodiment of math problems in their graceful curving. Farther upriver, just beyond the construction project, the bed is soft, with trees and brush growing in the margins, fish in the small pools, night herons and great blue herons and egrets eyeing the fish. Farther downstream, it's all concrete crossed by bridge after bridge. To the east, the Arroyo Seco splits hills, is environed by factories, warehouses, and industries of sundry stripe and hue. Bright blue bins and bright green trucks dot the land, sprinkled among the gravestone-grey buildings.

Signs of habitation on the plateau above the confluence include painted scribbles on the various structures (*Manny Was Here*) and odd ruins of things: standard-issue baseball pants and a Dodger cap, an unwound skein of something-or-other, a leopard-skin ball, bottles and cans and blankets and bags, scum and skitter. Leaves of scrubby plants stir in the unending wind. Down at the confluence, a white stake truck backs into the river from the access hole, lets another big truck into the stream, and pulls back out. Graffiti on the overlook bridge catches my eye:

LIFE
LIVES

The words stand out amid the freeway infrastructure and look like a caption for the confluence, a vague but insistent interpretation of vortex. A vortex is a ferociously spinning flow that brings together all manner of things, pulling them towards a center: a tornado gathering debris, for example, or fluids swirling down a drain. Figuratively, the confluence–vortex brings together the river's strange past with its dynamic present, the cultural history surrounding the Los Angeles River as well as the natural history of the stream. Price traces the shift from lifeline in a semi-arid region to concrete-lined flood control channel, dwelling on the conditions in what, for a while, was the Porciúncula, and in time became something other than a river. Angelenos stopped calling it a river, stopped thinking of it as a river, and eventually stopped thinking about it at all, allowing it to lapse from the lives they lived into the shadows of power's structures.

Many parties, beginning in the latter half of the twentieth century, have worked to "restore" or "revitalize" the river, to make it a river again. This suggests that cultural practices were capable of denaturing the river, of dismantling its nature and making it something else, and it further suggests that something can be done to re-nature the river. That line of thought, however, follows from the same separation of culture and nature that led to the river's neglect and demise. The apparent logic is straightforward: culture has the power to render nature null; culture also has the power to restore nature. This, however, is a limited way of seeing nature, one which does not reflect the complexity of nature or the effects of our conceptual framework in shaping what we take to be "nature" or "natural." What if, instead of un-rivering the river and then re-rivering it, we have all along been in the presence of a river? What if "the river" didn't go anywhere; rather, our actions and ways of thinking dried up and hardened, making us miss the river's nature? The seemingly simple statement *LIFE LIVES* suggests to me a strange analog: The River *Rivers*. Perhaps the Porciúncula/Los Angeles ever unfolds according to its nature as a river—a fact that our cultural practices have yet to catch up with.

Price's "Thirteen ways" experiments with possible means by which literature might help our minds play catch-up. Much like the tours she led for many years as an "urban ranger" in Los Angeles, Price's nonfiction does more than just point things out; it encourages rethinking of the concept of nature by exploring the stories that form and maintain different determinations of that concept's meaning. She uses the river to different ends, first to investigate some of the consequences of those stories, and second to begin to write a new story about nature in LA. Her essay proposes that we can begin to see the river differently, the city differently, and nature differently, only if we also re-examine the stories that historically and presently frame our vision. The goal of such writing is not

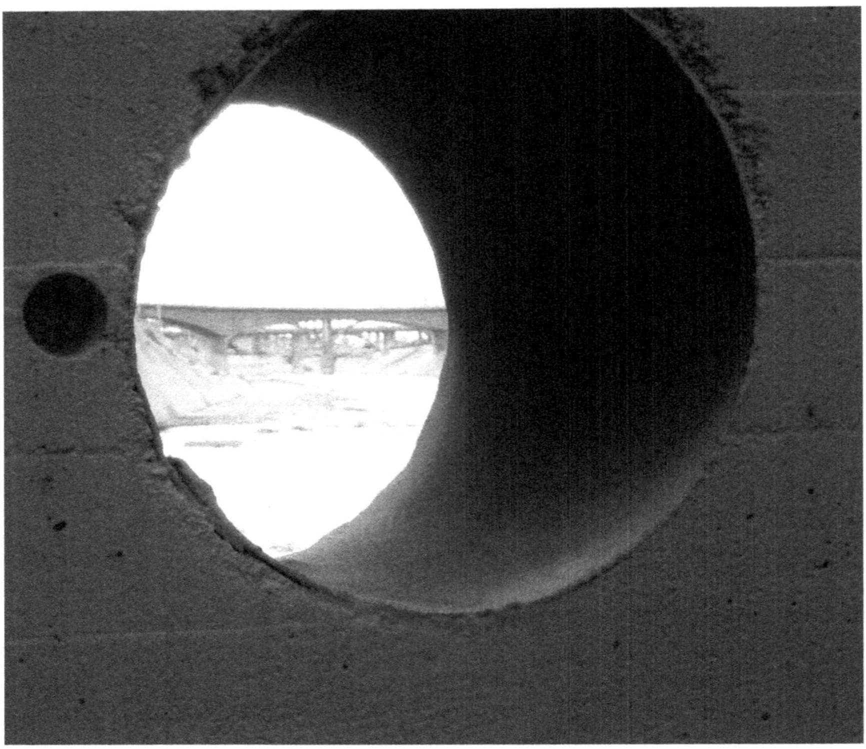

Figure 15.3 Confluence peephole

Source: Author.

to "get" nature; it does not require pinning down discrete objects, nor freezing some aspect of nature. While some forms of nonfiction operate by fixing a thing, an event, a fact in time or space, the object of study is thereby disconnected from the larger systems in which it circulates, its movement within those systems arrested. Meaning and truth are thereby diminished. Price experiments with another version of nature—a version at once more basic and more complicated.

The essay begins with a prologue that posits the river—"an outsize concrete sewer" that is "most famous for being forgotten"—as her favorite place to think about the city (Price, 2006, vol. 1, p. 2). Price acknowledges the river as an eyesore, but also as a site for seeing "nature" anew, and she focuses on "the Confluence," not far from Dodger Stadium and the place where the Spanish founded the pueblo, as a place from which a different way of writing about the nature of Los Angeles might emerge. Finding fault with the genre of "nature writing," especially recent iterations, Price observes that most of this sort of nonfiction is too caught up with ill-formed notions of the importance of wilderness—nature is somewhere "way out yonder," nowhere near the nation's urban centers. "Earnest, pious, and quite allergic to irony," many writers of nonfiction have unwittingly contributed to the cordoning off of "nature" from "culture," resulting in "one of our most pressing national cultural catastrophes" (Price, 2006, vol. 1, p. 3). Price tries to redress "the crisis in nature writing" through investigation of the different ways "nature" surfaces in the city.

The first half of the essay, while attentive to the peculiarities of Los Angeles, considers conceptions of nature that hold broader applications: "nature" as it exists in the wilderness and in the form of wild things, usually not present in cities (ways #1–2); "nature" as place (the "landscape and ecology we build in and manage"); and "nature" as resources (with which we "build cities and sustain our urban lives"), which carries different value, depending on location and access to resources as well as on "social position and point of view" (ways #3–5). Bringing together these different concepts of nature, Price raises a larger question regarding the separation of "nature" from "culture." She accuses most nature writers in American culture of propagating a myth of nature as "Wildness. Not-us-ness. The anti-modern. A place apart. Salvation. Refuge." This "ur-meaning" of nature continues to function as "an exceptionally powerful American cultural assumption," one that denies us a more accurate, more intimate, more complete—in all, more complex—understanding of the nature of "nature" (Price, 2006, vol. 1, pp.11–12).

These ideas belong to the sixth way of seeing nature: "As a Premier Source of Human Meaning." Stories about nature are fundamental to human experience and our efforts at making sense of the world we inhabit. In Price's view, recent and contemporary versions of nature have impoverished our understanding and contributed to poor policies. Better, more nuanced, more accurate and connective nature stories will enable us to develop fuller explanations for this place and help "guide [our] actions within it" (Price, 2006, vol. 1, p. 11). She summarizes her argument near the end of the essay's first half: "Our foundational nature stories should see and cherish our mundane, economic, utilitarian, daily encounters with nature…" (Price, 2006, vol.1, p. 13). For the most part, current foundational stories separate us from nature, and this separation structures the many systems by which we operate and the myriad institutions that maintain those systems. While great things have been accomplished, many practices have gone wrong—Price mentions the "wanton destructiveness of toxic industrialism and excessive consumerism" that have yielded political and financial inequality, environmental degradation, loss of resources, personal and public desperation. To deal with urgent problems, we need to be able to detach from those stories about nature in order to envision others. "Thirteen Ways," connecting disparate views and multiple parties, calls for a new kind of nonfiction that will participate in the revision of "nature" as we have come to know it. Nature, at once a source of meaning and a product of our methods for making meaning, is the question for which our modes of inquiry and possible responses are of utmost importance.

***

The arroyo seeps into the wide blank bed over some truck tire tracks, joining the rippling current of the central notch that bears the burden of the LA River's contents. Downriver, in the distance, the stream disappears beneath bridges, and on either side of the notch the wider bed dries under broken clouds. I cross bridge after bridge to see the river in segments, heading toward the ocean, looking up and down the course, reconnoitering the banks: North Broadway, North Spring Street, North Main Street, Cesar E. Chavez Avenue. From Broadway on the "Historic Buena Vista Viaduct," I can look back to the confluence ensconced below the bluff I climbed earlier. A tent stands tucked into the shadows by the bridge. The Cesar E. Chavez bears a dedication to Fray Juniper Serra, and the river swerves away from the setting sun toward the next set of bridges: First, Fourth, Sixth, Seventh, Olympic Streets. These downtown structures are massive viaducts, covering the river for just a small portion of their span with the bulk crossing rail yards and tracks and warehouses. Long walks up steep, feces-filled stairs and along sidewalks edged by vehicles take me to the rough center over the river. At Seventh, sixteen Black-necked Stilts stand on the upstream side of the bridge, the first birdlife I've seen for a while. At Olympic, a trash island has formed on the upstream side and a few Stilts gather on the downside. It is silent on the riverbed with industrial clamor all around:

Figure 15.4 Paying tribute

Source: Author.

terrible traffic, a building being wrecked, scrap being recycled, trains groaning and lurching. At the center of it all, beaten on by intermittent sun, glares the emptiness of the river.

Price concludes the first half of the essay with a section titled "River Trip #1," a conceptual excursion into the region's cultural history and the apparent misplacing of its namesake waterway. The section performs several significant functions: 1) it brings the river's role in the essay to front and center; 2) it provides a prime example of the nature–culture problems under discussion, extending the essay's humor, critical insight, and compassion to a particular case study; and 3) it positions the river as a kind of vortex around which numerous stories whirl. The river's appearance in "Thirteen Ways" mirrors its actual course through the basin and history. After it appears in the prologue, it disappears, surfaces, disappears, and ultimately re-emerges at the end, with the confluence depicted as the middle of the vortex, ushering in the potential for a new beginning. In the river's history, Price proposes, we can read "all of L.A.'s notorious troubles—environmental chaos, social inequities, community fragmentation, water shortages, water imperialism, and erasure of civic memory" (Price, 2006, vol. 1, p. 16). To bring that history to life, she shapes "Thirteen Ways" into a blend that makes use of different sub-genres of nonfiction, documenting but also going beyond documentation through imaginative association and informed interpretation.

Price practices the literary art she styles "nature writing" while developing theories about the genre, and her practice combines history and social criticism with humor and creativity. Writing about nature in LA means more than writing about "beaches, mountains, and canyons" (Price, 2006, vol. 1, p. 5). For Price it also involves mango body whip and the Beverly Center Mall; asthma rates in "the poorest, least white, and most industrial neighborhoods" (Price, 2006, vol. 1, p. 7); Zu-Zu the Chihuahua of Silver Lake and her struggles with a local coyote; Aaron Spelling's mansion in Bel Air; the urban demographics of "the second largest U.S. metropolis" (Price, 2006, vol. 2, p. 1); the disparity between the number of millionaires and the number of philanthropists; "rollerblading in a thong at the beach (whether male or female) while strumming a guitar" (Price, 2006, vol. 2, p. 7); gridlock and homelessness; Gore-Tex; the Clean Water Act; and park-space shortages (Price, 2006, vol. 2, pp.10–11). And most of all, nature writing in LA—writing that addresses the question, "How do we make nature meaningful?" (Price, 2006, vol. 1, p. 11)—involves the river.

The prevailing story about nature in LA was powerful enough to enable Angelenos to "lose track of an entire river" (Price, 2006, vol. 1, p. 13); to find the river, we need another story or another way of telling the story or another way of listening to the story. Price's essay gathers multiple elements of nature and multiple ways of making sense of those elements, the writing itself taking on some of the features of a vortex centered on the river. Historical scope, geographical range, critical insight, and a humorous style—these are the means by which she brings the life in and around the river to her nonfiction. The combination of diverse phenomena, competing stories, and multi-faceted interpretation comprises the vortex. And yet the vortex is more than a mere collection of things: it is also the method by which elements combine. It is not a "done deal" with definitive boundaries but a sort of inclusive, active, and ongoing "combining"—related, I think, to the cryptic graffiti I saw by the confluence: LIFE LIVES. Nonfiction that tries to get at the truth of life must necessarily be incomplete, must recognize its own participation in the living it attempts to capture and convey. In both content and style, "Thirteen Ways" expresses compassionate concern for our belonging to and our being in nature. To do so, the essay situates the river as a centerpiece of life living. Price's non-fictional vortex brings out the life *of* the river—an alternative approach to revitalizing, or restoring life *to,* the river. The river is alive. The river lives. With or without us, the river *rivers*, and our thinking could better embrace this fact of life.

★★★

From Culver City, my base of operations for these investigations, I drive the 10 to the 110, traffic rushing heavier than on previous days; near the confluence—experienced this morning as a rumor somewhere below a dizzying tangle of freeways—I change to the 5 and head upriver a few miles, exit onto Los Feliz Boulevard, and park at the par-3 golf course. The reach of river I am exploring today has a long stretch of soft bottom, which begins just below the river's great bend in Glendale and runs over a mile down to just beneath the Los Feliz Bridge. Though for most of the river's length its waters run over concrete bed, in several short sections one encounters "soft-bottom channel with annually flooded riparian growth" (Garrett, 1993, p. 5). Both the downstream and upstream ends of all the soft-bottom sections are graced by transition zones, with a distinct line separating the concrete bed from the sand, pebbles, and vegetation of earthen bottom where active underground springs prevented cementing. A transition zone features the junction of two distinctly different bottom-types: soft and hard, earthen and concrete, apparently natural and apparently (and, to most ways of thinking, "obviously") artificial. The surface flow wrinkles as the water moves from one bed type to the other, wavelengths loosed in symmetric succession. Birds gather amid the rocks and vegetation: pigeons, mallards, gulls, herons. A cormorant glides downstream low, a few inches above the water, and skids to a halt just before the break between beds.

Tiny reddish-brown birds flit about the grass plants growing on the concrete banks, which slope down to the water and flatten into a shelf pocked and puddled in places. There is life all around as I walk upstream, close to the water's eastern edge. Down here, I feel as if I've stumbled into an alternate reality of LA. A great blue heron stands still in midstream, water babbling past, a sound in stark contrast to the I-5 traffic in the background. The bird is as still as can be, thrillingly still. Egrets, coots, and wood ducks come and go. Occasional spray-painted testimonials express the trials of human experiences near the river. By the cover of the drain, inscribed in blue: "She's to [sic] good / to be true." Walkers and runners populate the path on this side of the river, bikers on the other side, birds everywhere. A green heron steps onto a rock in midstream, pauses, takes off, and then two other herons explode out of a tree on the opposite bank. Wigeons float past, a cinnamon teal; four cormorants roost in a dying tree; killdeer, plovers, and turnstones fuss, forage, and flutter away. I come to another outlet, a three-chamber drain converted into a makeshift camp, swept and orderly: tent in one section, some clothes in another, bag of food in the third, each bay separated from the other by a drain wall.

At the upper end of this soft-bottom section, the shelf at the base of the bank breaks off. I walk on a little farther on the slant of the hard, smooth bank, and stop to get my bearings. From here, looking north, I can see the concrete bed, empty of all but a shallow coat of water, widen into the great bend around to the west and the Verdugo Wash coming in from the east. To the south, birds of all sizes and colors—black, gray, blue, dark red, white—congregate on the downstream side of this transition zone. Beneath the towers of power cables that cross from bank to bank at an angle, I feel as if held by a magnet to this strange spot. An egret stands near by, west of a building wall that tops the bank and bears the word *QUIXOTE*. Behind me, cut into the side of the bank, another outlet shelters someone's scattered belongings, and one word has been painted with care on the cat-face shaped drain cover: *Please*.

The transition zones have a magical quality, perhaps because of the immediate quiet surrounded by city throb, or the abrupt shift from blank bed and emptiness of scene to biotic profusion and variegation. Down the hard-bottom stretch below the Verdugo Wash confluence, black water rides in wide and smooth, is regularly corrugated by the concrete, then, arriving at a break that spans the riverbed, the water tumbles into a froth. Now white with bubbles, the water braids around islands, rocks, bars. Birds proliferate. Held at the quixotic seam for how long I cannot say, I finally head up the slanted bank and walk south into the sun. Taking the levee-top path back, I'm easily distracted from the river by the human activity: workers, some in lab coats, out for a break-time walk, others jogging; industry and grounds-keeping churning on; gardens gleaming and stables astir; ever the freeway rumble, roar, and whine across the river and below the Griffith Park hillsides. Back near the Los Feliz Bridge, an osprey flies away from the river carrying a fish. Coming to another transition, I am again arrested by the contrasting reaches and try to read the line between—a notable marker of separation but also connection.

The river, an alternatingly absent and present center of Price's essay, is the place "where all the possible kinds of nature stories in L.A. converge" (Price, 2006, vol. 1, p. 6). Her work on that convergence—a whirling, forceful gathering up of disparate phenomena she labels "an American narrative vortex"—converts her own essay into a kind of vortex, a writing that sticks together that which we conceptually keep apart, stuff we tend to call "natural" drawn in along with stuff we tend to call "artificial," "cultural," "political." Setting these elements in motion together, Price tries to pull readers out of certain ways of seeing nature toward different views, in order to imagine nature *into* rather than separate from our lives (Price, 2006, vol. 1, p. 13). Like the fictional twister in *The Wizard of Oz*, Price's non-fictional vortex transports us from a cut-and-dry "nature," separated from "culture" and depicted in black and white, to a more complex, vivid, Technicolor version.

Figure 15.5 Transition zone

Source: Author.

The river functions in the essay as a guide to another way of seeing nature. "Thirteen Ways" proposes that even a lost river, an absent river, a dry or concrete river, has meaning—perhaps no less than a "present," pastoral, or wild river. In fact, Price's essay hints that even when seemingly absent, a river can be present. Los Angeles lost its river by not seeing it—hence the title's focus on different ways of seeing. The essay tries to make plain that certain conceptions of nature led to losing the river; to counter that apparent loss, "Thirteen Ways" substitutes the supposition that a river (emblematic of nature) is always there, even if Angelenos (emblematic of culture) don't always see it. At the end of the essay when Price first arrives at the confluence—a place she herself has trouble finding, on a river she herself had neglected—she is acting out a potentially important movement: not a nostalgic return to some lost, wild nature but a going forward, onward, to a nature always before us, around us, in us.

In that movement, "Thirteen Ways" demonstrates how non-fictional literature might contribute effectively to understanding rivers. By combining "natural" and "cultural" phenomena and questions, the essay-as-vortex goes beyond the customary functions of nonfiction as a transparent medium that delivers some sort of unadulterated truth about history, place, phenomena, events, and so on. Rather than separating elements from their context or the medium from the message, the vortical approach emphasizes connection. Price's nonfiction attempts to help us think about our stories—those we tell but also those we live by—as the path toward a truer, more complicated approach to meaning. Stories about nature are not just philosophical or theoretical problems; they are foundational questions that have material, political consequences. Understanding those stories can help us "recover" seemingly lost nature by helping us *un*cover what is always to be seen.

If we want to improve the nature of our rivers, we need to revitalize and restore our thinking—return thinking to life, restore thinking to nature (and "nature" to thinking). Such a recovery might help us to understand, with greater clarity, our connectedness to—our belonging to—the nature we try to improve. This is where nonfiction might come in. It is a form of writing that, while documenting aspects of the world, also enables us to see more clearly our thinking about the world, our ways of seeing nature. All nonfiction is, in some ways, indexical, trying to point something out: sometimes it points out more or less objective facts and truths, sometimes more or less subjective, sometimes the transition zones between objective and subjective. The pointing out helps us see something about our language, about our thinking, which in turn helps us see something about the nature of our world. In other words, the index, converted into vortex, helps us to reflect on our ways of thinking and the world such thinking produces.

The historic disregard for and neglect of the LA River has resulted in part from a disconnection between what we call "nature" and what we call "culture." Jenny Price's work urges us to redress this disconnect by reconsidering the stories in which we live and by which we make sense of the world around us. To begin to address the various problems resulting from that disconnection—from environmental damage to economic injustice—we need to see the links between ecosystem and economy, place and people, "nature" and "culture." If, for example, we single out the LA River as a discrete object with its own issues fixable with this or that adjustment, we leave in place the foundational concepts of division, neglect the myriad systems in which the river is involved, and fail to address the complex conditions that produce the problems we try to solve. Discussing her ongoing project in and around LA, Price observes that "You will never have healthy riparian ecosystems if you only clean up the creeks in affluent neighborhoods." To her, "It's less about an environmentalism that preserves wild places, and it's less about sustainable cities or social justice than it is about showing how those things are connected and how environmentalism should showcase those connections" (Carruth, 2014, para. 43.). The LA River, a study in complications and implications, can only be "restored" or "revitalized" through multi-faceted approaches that foster connection-making.

Lewis MacAdams returned my call to set up an interview, reaching me as I had turned off of the Riverside/Figueroa Bridge onto San Fernando, just as I drove past the confluence. We met later in the week near the FoLAR office at the Los Angeles River Center and Gardens and sat near a fountain, talking about rivers and writing, poetry and politics, life and love. A prolific writer of both poetry and prose, he tells me about the literary attempt to clear up language, to clean up language and "bring some clarity to the universe." These ideas are not unrelated to his four decades of political work on behalf of the river—both arenas entail making ideas material, bringing concepts to life and imagined outcomes to fruition. MacAdams likes to call himself an "infrastructuralist" instead of environmentalist; with FoLAR, he is "trying to build a better river." As with Price's essay, such an endeavor is not nostalgic; it requires reflection, imagination, and thoughtful participation in the present to create a vital, viable future. FoLAR, "a 40-year art project," began as a (poorly received) performance piece, part of which involved MacAdams carrying borrowed chain cutters to the banks, cutting the fence, and declaring the river "open." That symbolic act has had radically material results, and he speaks with some pride about the recently signed law (SB 1201) officially classifying the river as a river and granting greater public access. To MacAdams, the river hasn't changed as much as people's minds have. But in changing people's minds about the river, the river too changes. The confluence—of words and deed, of mind and river—is exactly the point.

Notes

1 In celebration of a feast day for St. Francis and a small church—La Porciúncula, "little portion"—associated with him.
2 An earlier version of the essay appears in the collection *Land of sunshine: An environmental history of metropolitan Los Angeles*, edited by Deverell and Hise (2005). Citations refer to the two-part version published by *The Believer* (April and May 2006).

Works cited

Carruth, A. (2014). The power of play in urban environmentalism: Interview with Jenny Price. *Resilience* 1:1.

Crespí, J. (2001). *A description of distant roads: Original journals of the first expedition into California, 1769–1770*. (Alan K. Brown, Trans.). San Diego, CA: San Diego University Press. (Original work published in 1771.)

Deverell, W. and Hise, G. (Eds.). (2005). *Land of sunshine: An environmental history of metropolitan Los Angeles*. Pittsburgh, PA: University of Pittsburgh Press.

Garrett, K. (1993). *The biota of the Los Angeles River: An overview of the historical and present plant and animal life of the Los Angeles River drainage*. Los Angeles, CA: Natural History Museum of Los Angeles County.

Gumprecht, B. (1999). *The Los Angeles River: Its life, death, and possible rebirth*. Baltimore, MA: Johns Hopkins University Press.

Price, J. (2006). Thirteen ways of seeing nature in L.A. (Vols. 1–2). *The Believer* (April and May).

Shannon1. (2011). Map of the Los Angeles River Basin, CA. Retrieved from https://commons.wikimedia.org/wiki/File:LARmap.jpg.

Index

Bold page numbers indicate figures, *italic* numbers indicate tables.

www.ingramcontent.com/pod-product-compliance
Ingram Content Group UK Ltd.
Pitfield, Milton Keynes, MK11 3LW, UK
UKHW011000180926
475701UK00004B/478
9780367227722